I0034163

Precoat Filtration

AWWA MANUAL M30

Second Edition

FOUNDED 1881

American Water Works Association

MANUAL OF WATER SUPPLY PRACTICES ó M30, Second Edition
Precoat Filtration

Editor: Phillip Murray
Project Manager: Kathleen A. Failer

Printed in the United States of America

American Water Works Association
6666 West Quincy Avenue
Denver, CO 80235

ISBN 0-89867-787-4

Printed on recycled paper

Contents

Foreword

The first edition of AWWA Manual M30, *Precoat Filtration,* published in 1988, was prepared by the Precoat Filtration Subcommittee of the Coagulation & Filtration Committee, AWWA Water Quality Division. The subcommittee consisted of Ray W. McIndoe, Chair; Gary S. Logsdon; James L. Ris; and Alan Wirsig.

Regulations stemming from the 1986 amendments to the Safe Drinking Water Act, specifically the Surface Water Treatment Rule, sparked a renewed interest in filtration techniques, including precoat filtration. The Coagulation & Filtration Committee was asked to review and update M30 in light of both the new regulations and recent experience with this process. A new Precoat Filtration Subcommittee was formed to accomplish this task. This subcommittee consisted of

C. Michael Elliott, Chair, Stearns & Wheler, Cazenovia, N.Y.

Catherine M. Spencer, Wright-Pierce Engineers, Topsham, Maine

Ray W. McIndoe, Plymouth, Mass.

Stephen G. Vanderbrook, Larsen Engineers, Rochester, N.Y.

George Attenboro, Village of Newark, N.Y.

The first task of the subcommittee was to review M30 and solicit input from experts in the precoat filtration area. The subcommittee distributed more than 40 copies of M30 for comments to

- Coagulation & Filtration Committee members
- water suppliers
- engineering firms
- regulatory agencies
- equipment manufacturers
- media suppliers
- universities

Though the names are far too numerous to mention here, the subcommittee's thanks and appreciation go to each person who spent time and effort reviewing the manual and providing comments and suggestions. Based on review results, the subcommittee elected to maintain the established format of M30, revising where appropriate.

The consensus is that precoat filtration is a viable and appropriate filtration technique where source waters are low in turbidity and color. This is true for smaller systems because of the ease of operation and the reduced complexity of residual disposal. Precoat filtration has also been demonstrated appropriate for larger systems, especially where recycling precoat media is applicable.

Based on our review, further research is required for precoat filtration to continue to economically meet regulations. Areas requiring additional research include

- media reuse for small systems
- pretreatment with microscreens, coagulant, powdered activated carbon, or anionic resins
- particle characteristics versus process efficiency
- organics removal and disinfection by-product regulations

*   *   *

The second edition of *Precoat Filtration* was reviewed and approved by the Coagulation & Filtration Committee (808) of AWWA's Water Quality Division. Members of the committee at the time of approval were

R.W. Bailey, Chair, CH2M Hill, Orlando, Fla.

C.F. Anderson Jr., City of Arlington, Arlington, Texas

H.H. Bryant, Black & Veatch Engineers, Aurora, Colo.

D.A. Cornwell, Environmental Engineering and Technology, Newport News, Va.

H.J. Dunn, Regional Water Authority, New Haven, Corm.

C.M. Elliott, Stearns & Wheler, Cazenovia, N.Y.

L.E. Elliott, Camp Dresser & McKee, Inc., Wichita, Kan.

G.M. Faustel, Albany, N.Y.

K.R. Fox, Drinking Water Research Division, US Environmental Protection Agency, Cincinnati, Ohio

C.A. Griffin Jr., Camp Dresser & McKee, Carlsbad, Calif.

D.J. Hiltebrand, Malcolm Pirnie, Inc., Yorktown, Va.

M.A. Hook, Tampa Water Department, Brandon, Fla.

G.J. Kirmeyer, Economic & Engineering Services, Bellevue, Wash.

M.F. Knudson, Portland Water Bureau, Portland, Ore.

B.W. Kuebler, Los Angeles Department of Water and Power, Los Angeles, Calif.

A.L. Lange, Concord, Calif.

C.B. Lind, General Chemical Corporation, Syracuse, N.Y.

N.E. McTigue, Environmental Engineering and Technology, Inc., Newport News, Va.

T. Myers, CH2M Hill, Milwaukee, Wis.

W.E. Neuman, American Waterworks Service Co., Voorhees, N.J.

N. Qureshi, Progressive Consulting Engineers, Minneapolis, Minn.

K.P. Rogenmuser, Roberts Filter Manufacturing, Darby, Penn.

C.S. Wilder, Camp Dresser & McKee, Inc., Atlanta, Ga.

J.S. Taylor, Division Liaison, University of Central Florida, Orlando, Fla.

Chapter 1

Introduction

Precoat filtration is a US Environmental Protection Agency (USEPA) accepted filtration technique for potable water treatment. When the Surface Water Treatment Rule (SWTR) went into effect in 1989, it provoked a renewed interest in this filtration process as well as other accepted filtration methods. This manual provides general guidelines for the use of precoat filtration, commonly called diatomaceous earth (DE or diatomite) filtration, in potable water treatment. It includes an evaluation of appropriate applications for precoat filtration, discusses the design of economical filtration units, and presents an overview of operating practices.

HISTORY

During World War II, the US Army needed a portable, efficient filter to remove *Entamoeba histolytica,* a protozoan parasite prevalent in the Pacific war zone, from drinking water. The army developed precoat filtration, which successfully removed the cysts. Since 1949, precoat filtration has been used in the filtration of sugar syrups, fruit juices, wine, beer, and water. More than 170 potable water treatment plants using precoat filtration have been constructed. New precoat filtration plants are anticipated as a result of the SWTR, including a 300-mgd plant for the treatment of New York City's Croton Reservoir supply.

DESCRIPTION

In precoat filtration, unclarified water containing foreign particles is forced, under pressure or by vacuum, through a uniform layer of filtering material (media) that has been deposited (precoated) on a septum. The septum is a permeable support for the media and is sustained by the rigid structure of the filter element. As the water passes through the filter media and septum, suspended particles about 2 μm and larger are captured and removed. Figure 1-1 presents a size spectrum of waterborne contaminants and various filter pores used in water treatment. The average pore sizes for DE are much smaller than the pore sizes of conventional filter media, such as sand.

1

Diameter, *m*

10^{-10}	10^{-9}	10^{-8}	10^{-7}	10^{-6}	10^{-5}	10^{-4}	10^{-3}	10^{-2}
1 Å	1 nm			1 μm			1 mm	

Particles

Molecules

Colloids

Suspended Particles

e.g.: Clays
FeOOH
SiO$_2$
CaCO$_3$

Bacteria

Algae

Viruses

Filter Types

Molecular Sieves

Filter Papers

Microsieves   Sieves

Membranes

Silica Gels

Diatomaceous Earth

Sand

Activated Carbon

Micropores   Pore Openings

Activated Carbon (Grains)

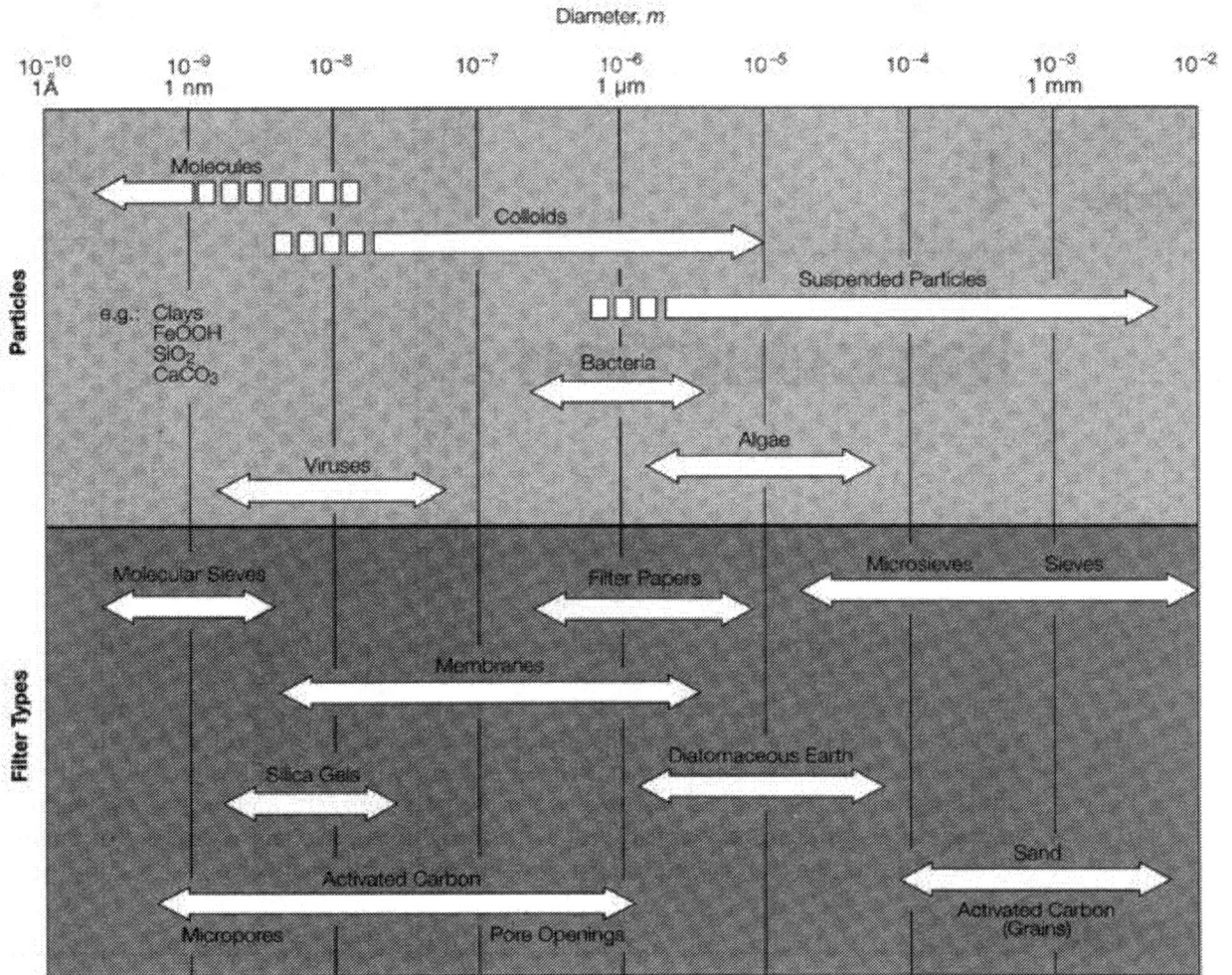

Source: Stumm, W. 1977. Chemical Interaction in Particle Separation. Environmental Science & Technology, *11:1066.*

Figure 1-1   Size spectrum of waterborne particles and pore sizes of filter media

The basic function performed by all water filters is to remove particulate matter from the water. Precoat filters accomplish this by physically straining the solids out of the water. Normally, there is no chemical reaction unless a soluble contaminant must be precipitated prior to filtration. The thickness of the initial layer of precoat filter media is normally 1/8–1/4 in., and the water passageways through this layer are so small and numerous that even very fine particles are trapped.

The majority of particles removed by the filter are trapped at the surface of the filter media layer, although some are retained within the filter layer. As water continues flowing through the filter, additional filter media, called body-feed, is regularly added to the incoming water flow in proportion to the particles being removed. The suspended particles intermingle with the body-feed particles to maintain flow through the filter. This mixture allows water to continue to flow through the filter as the accumulation of particles gradually grows thicker. The body-feed retards the head loss that might occur if foreign particles clogged the filter. Because the permeability of the filter cake (the mixture of precoat filter media, particles, and body-feed) is maintained, the length of the filter cycle is extended. However, as the filter grows thicker, a gradually increasing pressure drop through the filter system makes filtration impractical. When this stage is reached, the processes stopped, and the filter media and collected particulate are washed off the septum. A new precoat of filter media is then applied, and filtration resumes. A typical flow schematic for a pressure filter is shown in Figure 1-2, and Figure 1-3 shows a flow schematic for vacuum filters.

For small installations, the precoat tank functions for both slurry makeup and precoat recycle. Figure 1-2 illustrates the following procedure to precoat the septa:

1. Begin with a clean filter. Fill the associated piping between the precoat tank and the filter, as well as the precoat tank, with clean, filtered water.

2. Start the mixer on the precoat tank to create movement in the water. Add a batch of filter media that is sufficient to cover the septa in the filters, and mix to make the slurry.

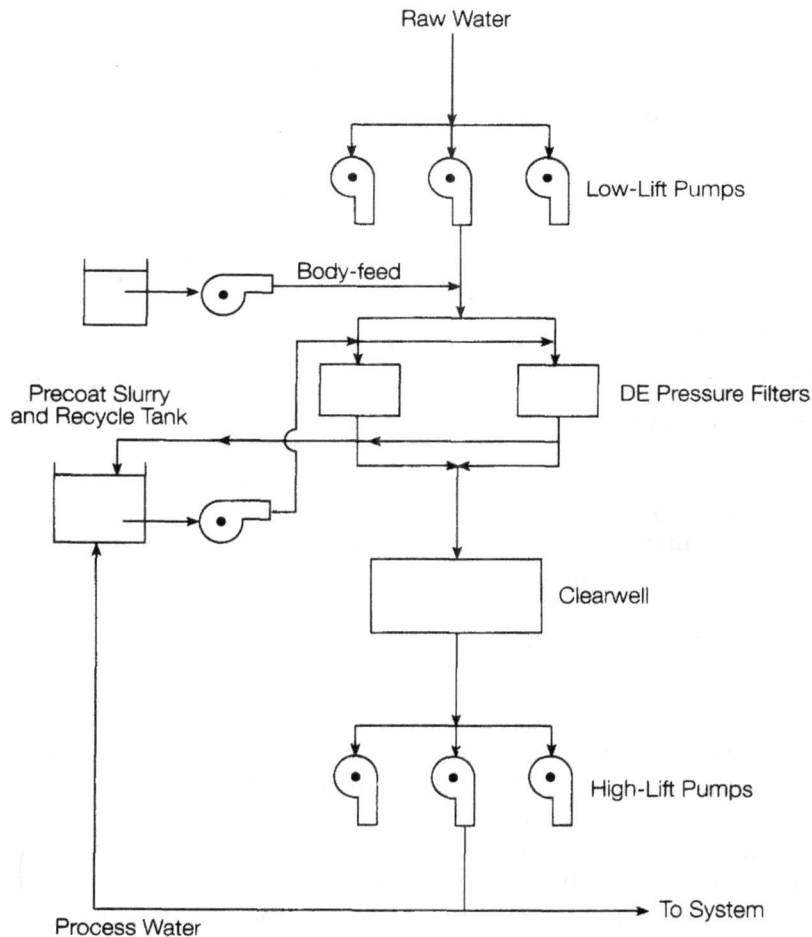

Courtesy of Stearns & Wheler.

Figure 1-2 Schematic of a precoat pressure filter system

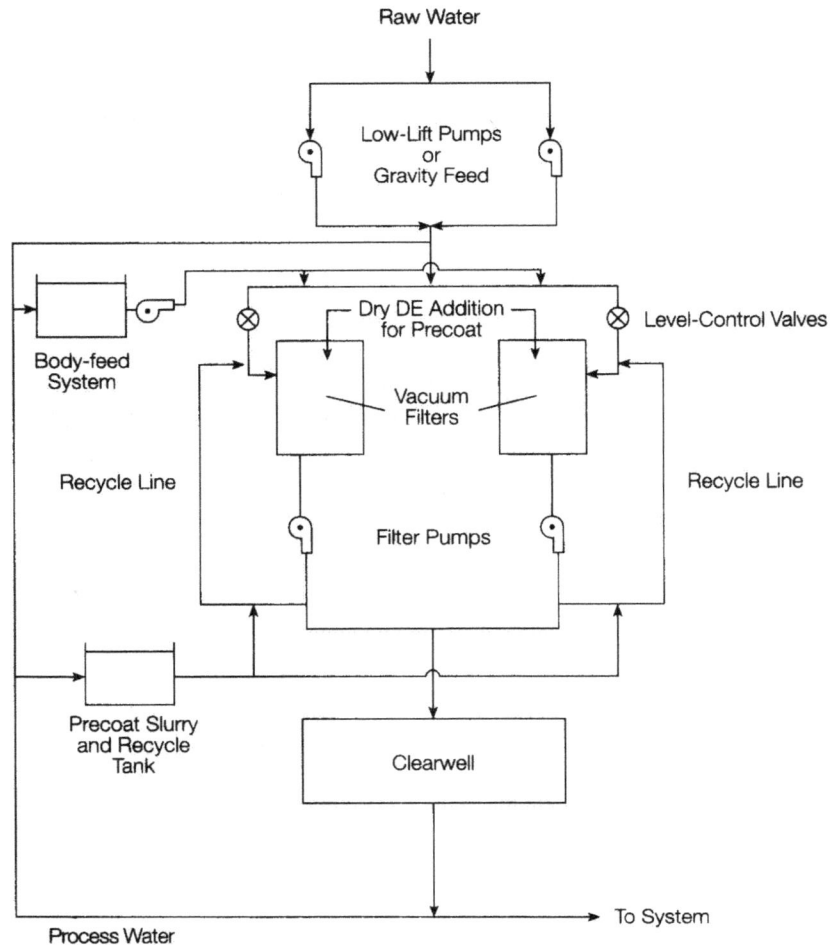

Courtesy of Ray W. McIndoe.

Figure 1-3 Schematic of a precoat vacuum filter system

3. Turn on the precoat pump to bring the slurry to the filter and to recycle the slurry and carrier water from the filter to the tank and back to the filter.

4. When all the filter media is on the filters, change the valving to bring raw water and body-feed to the filter. The filter septa have been covered with all the filter media when the turbidity of the carrier water is as low as the normal, clean filtered water. The filter is then ready for use.

There are five or six commonly used grades of filter media (sometimes called filter aid) available. Various grades of filter media perform differently in the way they remove particles and in flow characteristics. With an appropriate selection from among these grades, a large number of particles as small as 1 μm can be removed by the filter cake. This includes most surface water impurities. However, where small to mid-size particles (10^{-4} to 10^{-6} m) are present, filtration alone may not be adequate to reduce turbidity (particles suspended in the water) below the 1 ntu required by current regulations.

Generally speaking, precoat filtration is most effective when source water turbidity is moderate to low (10 ntu or less). However, higher turbidity levels may be

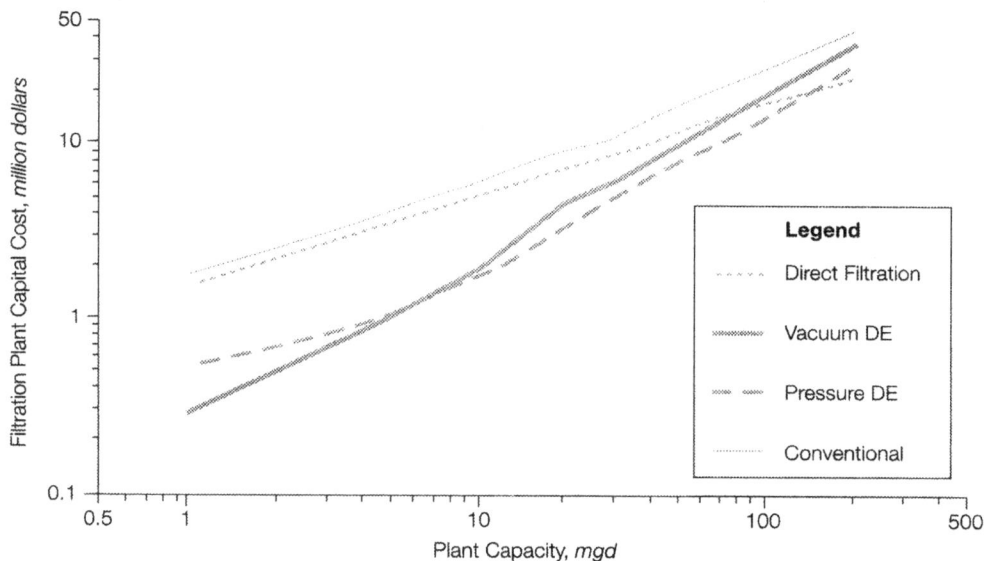

Courtesy of C.M. Spencer, Wright-Pierce Engineers.

Figure 1-4   Filtration capital costs versus plant capacity for various filtration systems

handled economically, depending on the concentration of the solids to be removed and on their physical characteristics.

Economic Benefits

Where the source water and other conditions are suitable, precoat filtration offers a number of economic benefits to the user, including the following:

- Capital cost savings may be possible because of smaller land and building requirements and lower installed costs (see Figure 1-4).

- Treatment costs may be slightly less than for conventional coagulation, sedimentation, and granular media filtration when filterable solids are low.

- The use of chemicals associated with granular media filtration, such as aluminum sulfate, iron salts, and polymers, is not necessary. The process is entirely physical and does not require operator expertise in water chemistry relating to coagulation.

- The volume of filtered water used for cleaning the filter is less than the volume used to clean granular media filters. Normally, less than 1 percent of the total filtered water is required.

- Diatomaceous earth filter residuals are easily dewatered, and in some cases, the media may be reclaimed for other uses, including soil conditioning and land reclamation. Research is under way to determine the feasibility of reusing filter media as body-feed.

M30

Chapter **2**

Applications and Economic Considerations

A number of factors must be considered when selecting effective, economical treatment and filtration processes. The most important factors are the quality of the source water coming into the filter and the desired quality of finished water. Source water quality can vary widely, depending on the source and type of water. Finished water quality may also vary, depending on its intended use and on local contaminants that must be removed.

Selecting a filtration process involves the following steps:

- Each proposed application should be examined to determine whether it produces a satisfactorily treated water, with or without other treatment steps.
- The process should then be compared with alternative types of treatments to see whether it is economically advantageous.

SOURCE WATER QUALITY CONSIDERATIONS

As a general rule, relatively clean source waters are good candidates for precoat filtration. Groundwaters and large impoundments of surface water from which only small amounts of suspended solids must be removed may be more economical to treat by precoat filtration than by conventional treatment. Designers should review historical water quality parameters before selecting or designing precoat filtration facilities. Some of the most important water quality parameters are discussed below.

Turbidity

Diatomaceous earth (DE or diatomite) filtration can be defined as a filtration technique for relatively low turbidity waters, generally below 10 turbidity units (ntu). A survey of direct filtration practice, including precoat filtration plants, indicated that 80 percent of the plants had an average source water turbidity of 5 ntu or less. The maximum turbidity at 80 percent of the plants was 40 ntu or less.

7

Table 2-1   Effect of pore size on flow rate and clarity

Precoat Media Grades	Average Particle Size,* μm	Median Pore Size,* μm	Flow Rate	Clarity
Fine	15	3.5	Lowest	Highest
↑	16	5.0	↑	↑
↓	22	7.0	↓	↓
	24	10.0		
	34	13.0		
Coarse	36	17.0	Highest	Lowest

Source: *Stearns and Wheler* and *Ray W. McIndoe.*
NOTE: Grades most commonly used for water filtration produce precoat cakes with pore sizes ranging from 5.0 to 17.0 μm.
*Typical sizes may vary slightly from one manufacturer to another.

Engineers should evaluate the efficiency and economics of pretreating source waters with turbidity levels higher than 10 ntu to obtain an influent with less than 10 ntu. Because turbidity is an indirect measure of the number and nature of the particulate in a source water, identification of the specific particulate present in a source water is recommended. Figure 1-1 portrays the size spectrum of waterborne particles of interest in contrast with the pore sizes of various filter media.

Both the number and physical nature of the solids present in source water affect the ability of a filter process to remove those solids. Some particulate, such as sand grains, some clays, and protozoan cysts, are nondeformable, discrete particles and do not pose problems. Deformable particles, however, tend to clog the media. In addition, some bacteria and viruses may pass through the filter. Turbidity readings, weight measurements of suspended solids, and particle counts and size distribution analyses are helpful techniques for determining precoat filtration's feasibility. However, those techniques alone may not be sufficient indicators of whether the process can be applied successfully and economically. Study of the types and number of algae present throughout the year and detection of the presence of organic colloids or fine clays in a water source proposed for precoat filtration are important in the evaluation process and in identifying the need for pretreatment.

Surface Water

Surface water sources must be evaluated on the basis of normal conditions that may prevail for most of the year. High-flow and runoff periods should also be examined for increased suspended material and its effect on the filtration process. Facilities for source water or finished water storage to provide adequate potable water supply during periods of adverse source water quality may be needed. Presetting of source water at the plant site may diminish some of the benefits of precoat filtration because of the increased capital and operating costs of the presetting facilities.

Many surface water supplies, such as lakes or ponds, may have algae, color, or taste and odor problems that require additional treatment in conjunction with precoat filtration. A microstrainer used before the precoat filter has been effective in removing microscopic material, including planktonic organisms and amorphous matter. This form of pretreatment has resulted in increased filtration run length.

Research has shown that precoat filtration with DE is capable of removing virtually all *Giardia* cysts normally encountered in low turbidity (<10 ntu) natural waters provided that the appropriate filter media is used and filters are properly

Table 2-2 Effect of DE particle and pore size on particle removal

Median Diatomite Size, μm	Median Diatomite Pore Size, μm	Particle Size, μm	Removal Rates, %
14	2.5	1	99.9
16.4	5	1	85
24	9	3	99.9
36	17	8	99.9

Source: C.M. Spencer, Wright-Pierce Engineers.

maintained and operated. To date, the removal efficiency of *Cryptosporidium* has not been well documented. However, recent experience with removal of particles and viruses from the New York City Croton Reservoir supply indicate the potential for good removal exists with this filtration technique. The following table indicates relative sizes of *Giardia, Cryptosporidium,* and coliform microorganisms.

Microorganism	Size, μm
Giardia	6–15
Cryptosporidium	3–6
Coliform	0.5–1

Although filtration generally is more effective with organisms larger than *Giardia,* the resistance of *Giardia* and *Cryptosporidium* to chlorine disinfection makes these organisms of particular concern. Table 2-2 indicates the relative removal rates of DE used in precoat filtration for particles in the size range 1–8 μm (Spencer 1991).

In a study by Schuler and Ghosh (1990), when filter influent between 0.1 and 1.0 ntu was spiked with 2,000 *Cryptosporidium* and *Giardia* organisms per litre at a DE precoat of 1 kg/m^2, the removal of *Cryptosporidium* oocysts exceeded 99.9 percent, with the addition of alum further improving removal. Conditioning the DE with a coating of aluminum hydroxide, producing a filter media sometimes called coated DE, may also enhance removal of 1-μm particles. When *Giardia* cysts are to be removed by precoat filters, a precoat layer of at least 1/8-in. depth or 1 kg/m^2 plus body-feed is recommended (Logsdon et al. 1981). Media particle and pore sizes, feed and flow rates, depth of precoat layer, and influent turbidity are some of the factors influencing particle removal. See chapter 4 for more information on filter media.

Groundwater

Groundwater supplies may require filtration to remove suspended material. In some cases, mineral impurities, such as soluble iron or manganese, require pretreatment to precipitate them from the source water. The resulting precipitate can then be removed by the precoat filter. (See the section on supplementary treatment practices, page 10.)

Pilot Testing

Where there is no prior practical experience, pilot testing of the water to be filtered is recommended before making a final process selection. Pilot testing the precoat filtration process involves using small-scale equipment similar in function and operation to the proposed full-size units. Pilot filters and ancillary supporting equipment are usually available from equipment manufacturers and producers of filter media. A pilot operation simulating the operating conditions required of a full-scale plant will help determine process applicability, design criteria, and operating economics. Pilot testing also familiarizes everyone involved with the workings of the filter system being studied. An involved and interested staff is critical to the success of a filtration plant.

Design engineers should check with state regulatory agencies having jurisdiction over public water supplies to identify operational and treatment efficiency parameters of concern for pilot test studies. States generally require a measure of the precoat filter's capability to remove *Giardia lamblia* cysts or particles in the size range of such cysts. The Surface Water Treatment Rule requires filtration processes to remove 99 percent (2 log) of *Giardia* cysts.

Supplementary Treatment

Additional treatment techniques can be used in conjunction with precoat filtration to ensure disinfection and to handle special problems, such as soluble iron and manganese, softening, color, taste, and odor.

Disinfection. Precoat filtration generally provides excellent clarification, and tests with the finest grades of filter media have shown good to excellent removal of turbidity and coliform bacteria. However, final disinfection with chlorine or other acceptable disinfectants must be provided, consistent with the multiple-barrier concept of public health protection.

Iron. Iron can be precipitated or separated from solution by mixing magnesite (MgO) with the source water, as well as with the body-feed, in a 10- to 15-min contact tank to provide sufficient reaction time. The resulting particulate matter is easily removed by the precoat filter. Also, iron could be separated by preaeration or other oxidation methods that make use of detention. Then the iron particles can be removed by the precoat filter. Tests have shown that solids remaining from the magnesite process are much easier to remove by precoat filtration than are the ferric oxide particles from the preaeration process. However, there may be cases in which a combination of aeration and the magnesite process may be more cost-effective.

Manganese. When manganese alone is the principal contaminant to be removed, potassium permanganate may be used to oxidize and separate the manganese. A detention time of up to 30 min at pH >7.5 may be required ahead of filtration, although there are plants that remove manganese by filtration after a 10-min detention period.

Iron and Manganese. Where iron and manganese are both present, there are several treatment options possible. Iron oxidizes more easily than manganese, so potassium permanganate will precipitate the iron first and then react with manganese. Stronger oxidants can be used.

Hardness. Where reduction of total hardness by lime and soda ash is required, residual calcium carbonate and magnesium hydroxide particles in the overflow effluent from the clarifiers have been removed with diatomite media.

Taste and Odor. Where taste and odor are problems, they may be handled by activated carbon in conjunction with precoat filtration. Preceding granular carbon columns with precoat filters will reduce the particulate load on the carbon beds, prevent the carbon media from clogging, and lengthen bed life. Also, the precoat filters may be evaluated to incorporate powdered activated carbon, along with the filter media, as precoat and body-feed, to achieve clarification and taste and odor removal within the filter.

ECONOMIC CONSIDERATIONS

After precoat filtration and appropriate supplementary processes have been shown to produce an acceptable finished water, several design factors must be considered in preparing design and specification documents. Because all these factors will influence the total capital and operating costs of the system, it is important to evaluate them individually and in relation to one another so that the final design will provide water most economically.

The following general economic considerations should be kept in mind when applying the more detailed design and operating guidelines that form the remainder of this manual.

Filtration Rate

The design filtration rate is expressed as gallons per minute per square foot (gpm/ft^2) of septum area and is sometimes called the surface loading rate. The higher the design filtration rate allowed, the smaller the required filter surface area to produce the total volume of water needed to satisfy demand. Decreasing the filter area will lower the initial capital costs. With less filter area, less precoat material is required for each filter cycle. On the other hand, higher filtration rates generally result in more rapid filter pressure loss during a filter cycle. This will result in shorter cycles with more frequent precoating. Some states have guidelines for filtration rates and other operational factors that affect design. Demonstration of filtration effectiveness and economy outside of the recommended filtration rates is usually required before state regulatory agencies will approve the design. A minimum suggested filtration rate is 0.1 gpm/ft^2 to keep the precoat on the septum. Filtration rates of up to 4 gpm/ft^2 have proven acceptable for certain installations. Most DE filtration plants presently operate within the range of 1–3 gpm/ft^2. Demonstration of successful operation at flow rates greater than 1 gpm/ft^2 may justify the use of smaller filter vessels, which afford greater economic benefit to the utility. These higher flow rates may also extend the permissible operating range of an existing plant so that the need for future expansion is delayed.

Terminal Pressure Differential

The terminal pressure differential is the maximum difference in pressure between the inlet to the filter and the outlet of the filter. Higher terminal pressure differential extends both the length of the filter cycle and the useful life of the precoat media. In other words, the longer the filter cycle, the more finished water that is produced per pound of precoat material used. However, inlet feed pumps used to provide the additional pumping pressure needed for the higher terminal pressure differential will increase power costs.

If variable-speed drives are to be used, pumping costs will vary throughout the filter cycle. This variation could affect operating cost estimates. Early in the design phase, the potential use of variable-speed drives on inlet feed pumps should be evaluated with regard to the overall filter run length and terminal pressure differential.

Grade of Filter Media

A coarser grade of media can be used to obtain longer filter cycles, as long as the process still produces acceptable finished water quality. Using a coarser grade of media results in a greater volume of water produced per pound of filter media. This gain may be offset by the added cost (20–30 percent) of the coarser grades of media

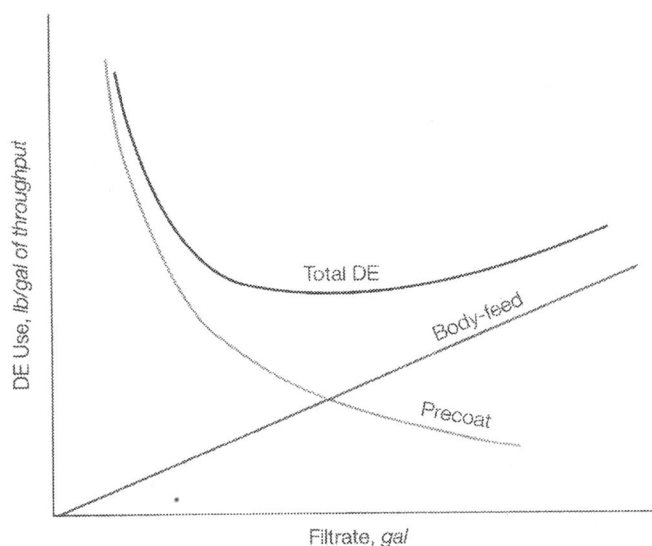

NOTE: Relationship is typical. Exact figures will vary from water source to water source.

Figure 2-1   Effect of body-feed on total DE use

relative to the finer grades. Present filtration using DE throughout the available size range has been proven to remove more than 99 percent (2 log) of influent *Giardia lamblia* cysts. However, removal of turbidity, heterotrophic bacteria, and total coliform bacteria is strongly influenced by the grade of DE used. Testing of the proposed water source is necessary to determine the precoat grade most suitable for that site.

Body-feed Rate

The body-feed rate is defined as the amount of media that is added to each gallon of source water during filtration to ensure that the filter cake remains porous to water flow. In general, the higher the body-feed rate, the more porous the cake, and therefore the longer the filter cycle. In some cases, however, excessive amounts of body-feed can increase cake thickness and introduce additional resistance to flow, which will shorten the cycle. It is important to analyze the body-feed rate in relation to filter run length to see whether an extra body-feed amount more than offsets the prorated cost of precoat material initially applied. Table 2-1 and Figures 2-1 and 2-2 depict typical interrelationships between these factors. The ratios at the top of Figure 2-2 indicate the relationship between pressure drop and the resulting increase in cycle run length.

As discussed previously in this chapter, both the amount and physical nature of the particulate to be removed affect filter operation, especially the selection of appropriate body-feed rates. For most applications, testing to determine the optimal body-feed rate (in milligrams per litre of body-feed per nephelometric turbidity unit of the source water) will be necessary.

Mathematical models have been developed to predict head loss as a function of filtration rate, water viscosity, septum shape, and body-feed use. For example, one computer program assists in designing new plants and optimizing operation of new and existing plants.

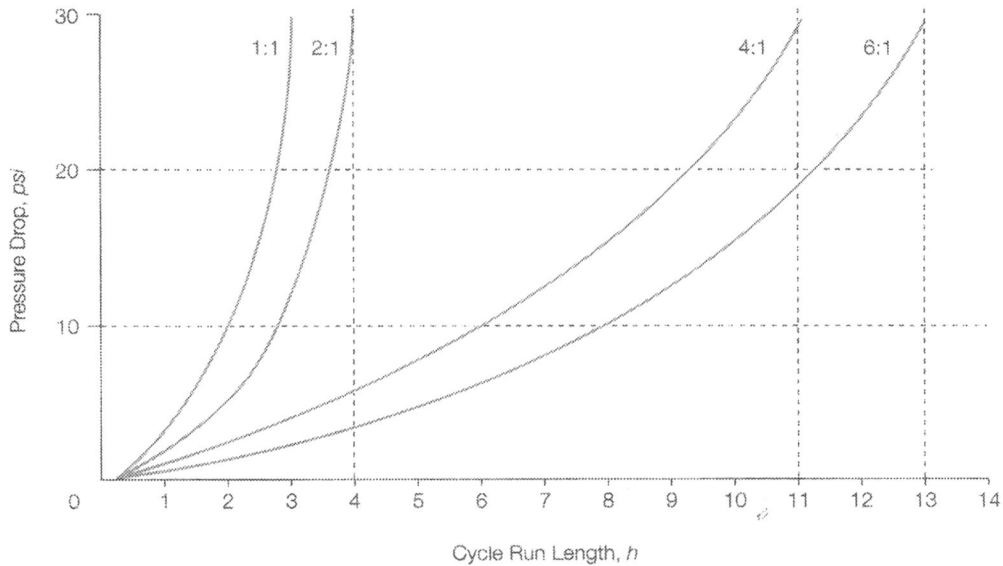

Figure 2-2 Cycle run length as a function of body-feed for DE plants

Operating Costs

When facilities are in place, the major variables affecting operating costs are filter media and pumping costs. Labor costs must also be taken into account in the overall design, because the cost of labor becomes one of the higher operating costs for a plant after capital costs have been paid.

Filter media and pumping costs. The target objective of the operating plant is to deliver the maximum volume of finished water to the distribution system at the lowest combination of filter media and pumping costs. Even though pilot tests may have been done, a series of simple tests conducted with the full-scale plant equipment helps to establish optimal operating routines. Under controlled conditions, the flow rate and body-feed rate are varied, and the resulting effects on pressure rise, filter run length, and turbidity are noted. Filter media use, including both precoat and body-feed, should be carefully observed and analyzed for the lowest-cost compromise of all factors. These factors must be correlated with the influent source water quality to allow for seasonal variations.

Material-handling costs. In addition to the media and power costs associated with filtration, there are added costs of receiving and storing filter media and moving material into the process stream, as well as costs associated with disposal of water and filter media waste slurries.

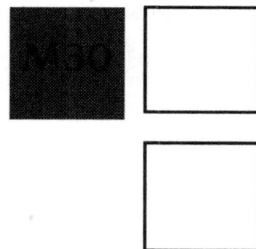

Chapter **3**

Filter Design

Several design decisions are important in ensuring effective, economical operation of precoat filtration systems, including

- type of filter to be used (pressure or vacuum)
- design of the filter element and septum
- hydraulics of the system
- rate of filtration

FILTER VESSEL DESIGN

As shown in Figures 1-2 and 1-3 (pages 3 and 4), there are two basic types of precoat filters: vacuum and pressure. The construction features of the vessel can be specified after a design decision is made to use one of these types.

Vacuum Filters

A vacuum filter consists of an open tank containing the vacuum filter elements and their septa as illustrated in Figure 3-1, Vacuum filter tanks and components are available in corrosion-resistant materials, such as fiberglass-reinforced plastics and stainless steel.

Suction created by a filter discharge pump or vacuum discharge leg downstream of the filter enables available atmospheric pressure to move water through the filter media cake as the cake builds up. During filtration and cleaning, the open filter tank permits easy observation of the condition of the septa, the elements, and the general condition of the filter (see Figure 3-2).

The maximum pressure loss should not normally exceed 7.4–8.4 psi, which limits the length of filter cycles when compared with pressure filters. However, vacuum filters are still capable of providing economically acceptable operation. When a vacuum filter is used, any pressure of water entering the plant must be dissipated; the pressure cannot be used for filter operation.

Air or dissolved gases that come out of solution because of a decrease in pressure across the filter cake and septum and are transported by the water (outgassing) may have an adverse effect on the filter cake. Outgassing tends to disrupt the integrity of

15

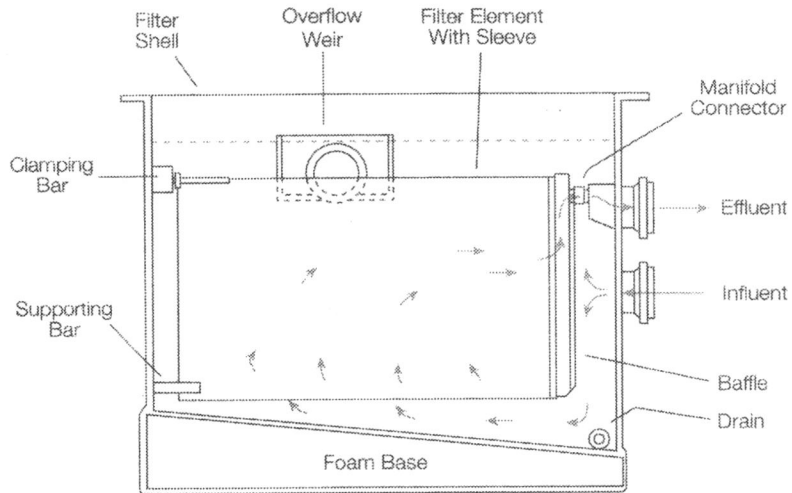

Courtesy of Westfall Manufacturing Company.

Figure 3-1 Vacuum diatomite (DE) filter element flow diagram

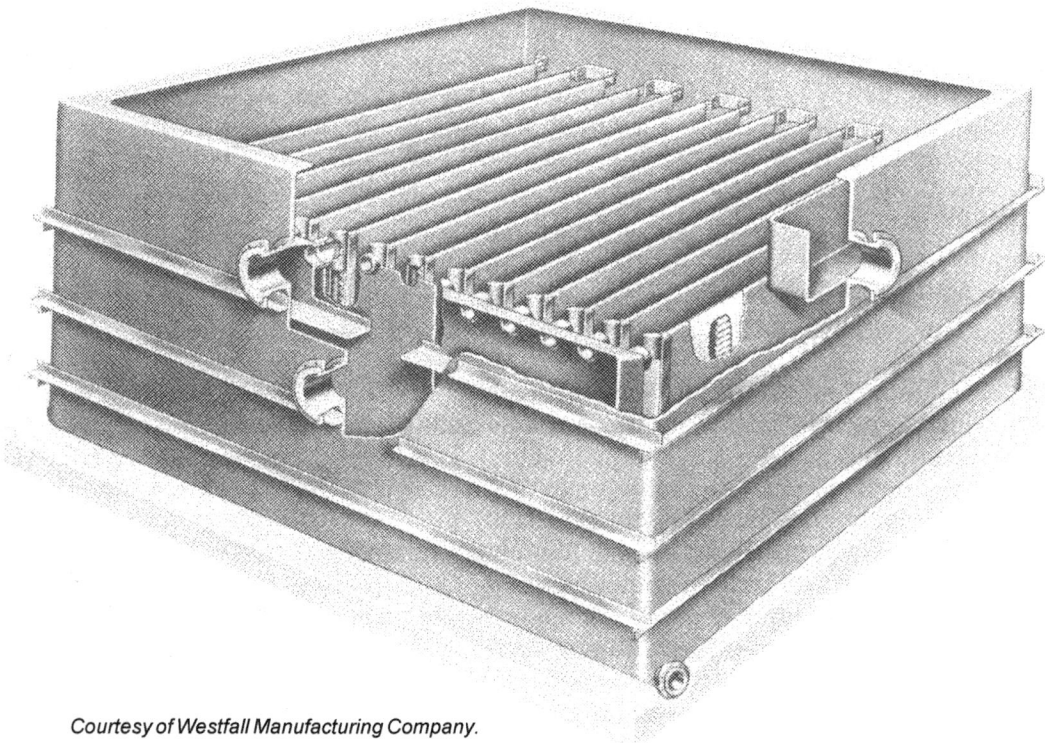

Courtesy of Westfall Manufacturing Company.

Figure 3-2 Vacuum diatomite filter

the filter media on the septum. Special care should be taken to eliminate any air being pulled into source water through pump glands or nonflooded filter-media slurry eductors. The amount of gas coming out of solution is directly related to the dissolved gas concentration, water temperature, and vacuum.

Pressure Filters

In pressure filters, a filter feed pump or influent gravity flow produces higher-than-atmospheric pressure on the inlet (upstream) side of the filter, forcing liquid through the filter media cake. Large pressure drops across the filter are theoretically possible (limited only by the strength of the filter shell and the filter elements and septa), but the maximum economic differential pressure drop is generally limited to 30–40 psi. Typically, a higher pressure drop across a pressure filter will yield longer cycles and will remove more suspended solids per pound of filter media than with vacuum filters. Increased pumping costs for differential pressures much greater than 30–40 psi, however, usually offset savings in filter media costs.

Several types of pressure filtration vessels have been developed. Figures 3-3 through 3-6 depict some of the possible configurations. Space considerations may favor the selection of one filter vessel configuration over others for a particular application. Any vessel selected should meet the following construction criteria. Pressure filter housings should be made of corrosion-resistant materials or should be lined or coated with protective materials that will last for many years. American Water Works Association (AWWA) standards for coating and lining tanks and water pipelines should be used as guidelines when specifying the coating and lining for precoat filtration facilities. Stainless steel should be considered for maximum service life. Vessels should be fabricated to ensure good construction and adequate strength in accordance with the American Society of Mechanical Engineers' Boiler and

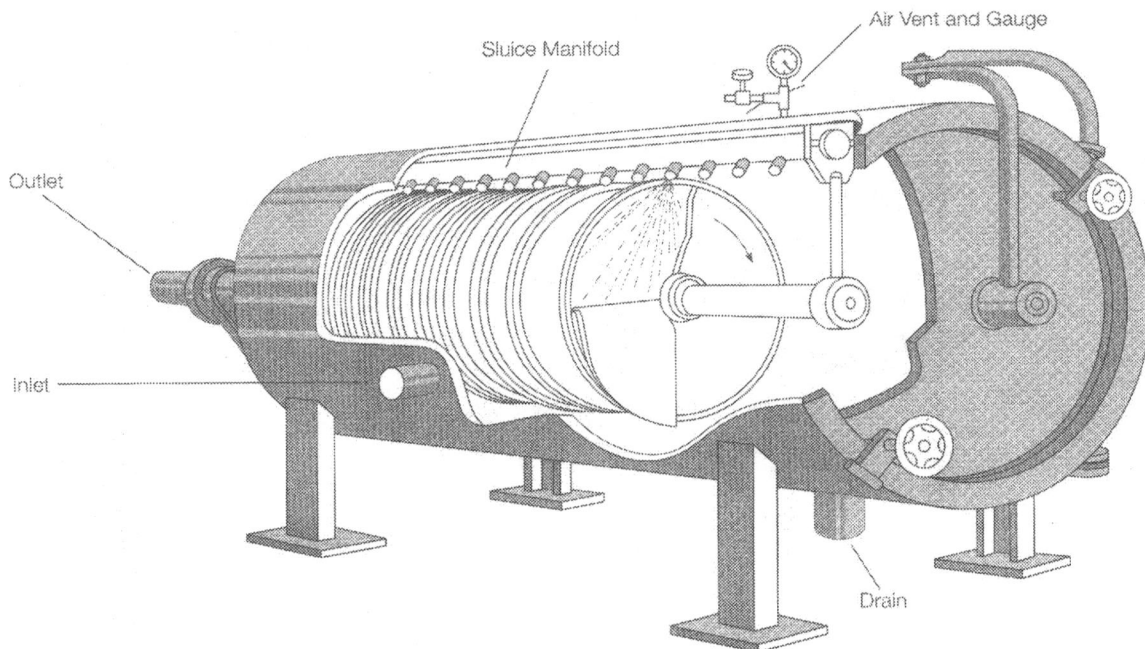

Reprinted with permission of Celite Corporation.

Figure 3-3 Horizontal tank pressure leaf filter—rotating leaf

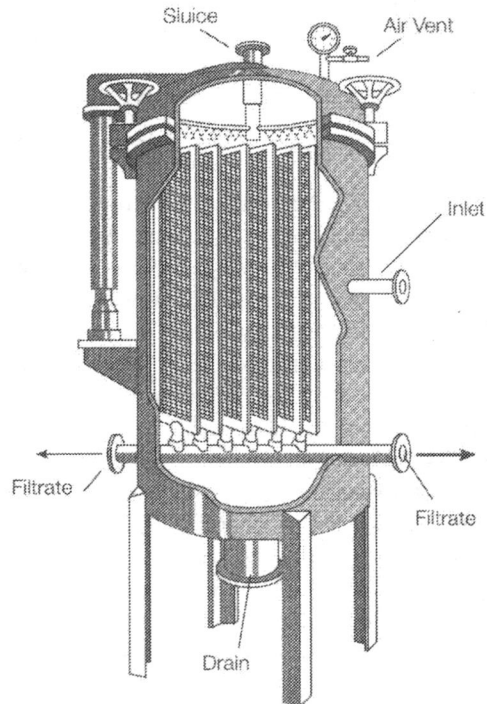

Figure 3-4   Vertical tank pressure leaf filter

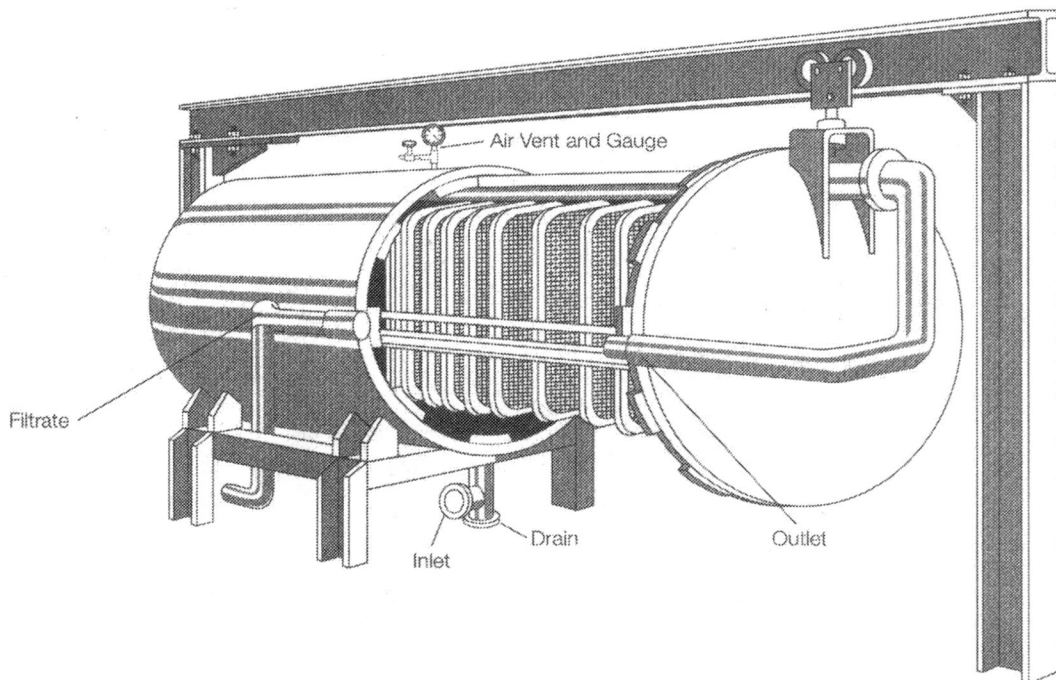

Figure 3-5   Horizontal tank pressure filter—vertical leaf

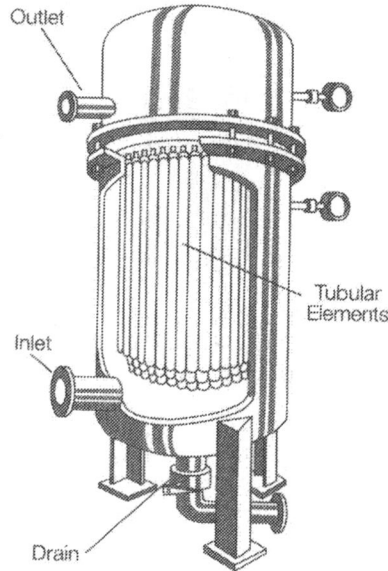

Reprinted with permission of Celite Corporation.

Figure 3-6 Tubular pressure filter

Pressure Vessel Code, Section VIII, or other guidelines. High-pressure relief should be provided. Vessel design should provide for full access for inspection and maintenance operations. One or more illuminated sight ports should be installed in the shell to permit visual observation of the filter during operation.

Access to internal areas of the filters may be achieved through quick-opening heads that allow the vessel to be opened within minutes (see Figures 3-3 and 3-4). Pressure vessels are also designed with fixed heads and retracting shells to permit immediate access to all of the septa (see Figure 3-5). Before inspection, pressure tubular filters generally require bolt removal and some piping disassembly, as well as a hoisting facility to remove the filter head and the tube sheet (see Figure 3-6). Suitable lifting devices should be provided, either over or alongside filters, to assist in the removal and disassembly of internal elements of the filter for maintenance. Piping, wiring, and auxiliary equipment should be arranged to facilitate normal operations and maintenance work.

FILTER ELEMENT DESIGN

The following basic criteria must be incorporated into element design:
- firm support for the filter media cake
- adequate drainage area inside the element so that the filtered water can easily exit from the element
- proper construction of the septum to provide clear openings of proper size so that the filter media forms strong, stable "bridges" over the openings
- capability of septum material to maintain the integrity of the weave pattern to prevent distortion of opening size or shape with continued use
- corrosion-resistant construction materials

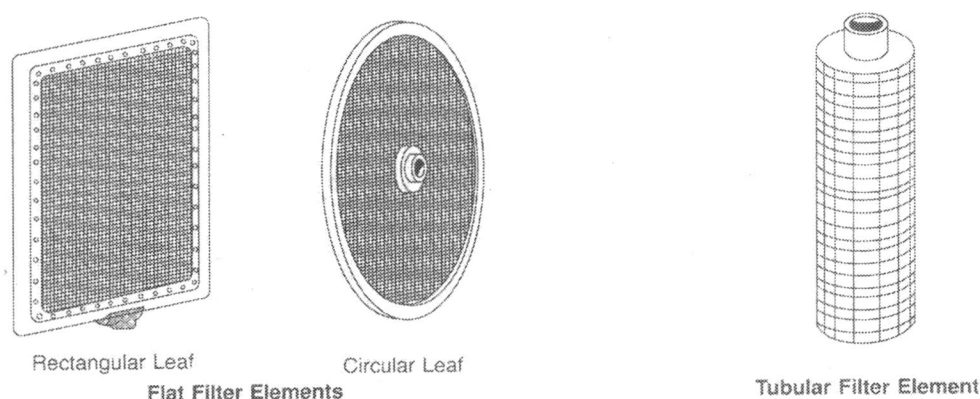

Rectangular Leaf   Circular Leaf
Flat Filter Elements   Tubular Filter Element

Figure 3-7   Typical filter elements

Filter elements may be either flat or tubular (Figure 3-7). Flat elements, often referred to as leaves, may be rectangular or round. Tubular elements are available in several different cross-sectional shapes but are generally round. Tubular elements, which are always oriented vertically in vertical tanks, require sufficient headroom for disassembly and routine maintenance, Flat leaf elements used in pressure filters are generally mounted vertically in horizontal or vertical tanks and discharge filtered water through a bottom nozzle into the filter discharge manifold. Flat leaf elements in vacuum filters are also generally mounted vertically, but they discharge filtered water through a top nozzle into a discharge manifold located near the top of the filter box. This arrangement permits escape of any displaced air or gas.

The internal construction of flat leaf elements can vary widely, depending on the intended service, but elements intended for water service will usually incorporate a drainage member to provide strength and rigidity as well as free drainage area (Figure 3-8). The septum material overlays the drainage member. It usually consists of either tightly woven stainless-steel wire mesh or a tightly fitted bag made from monofilament polypropylene weave. A frame member around the perimeter of the leaf provides additional drainage as well as strength (Figure 3-9). Tubular element construction varies widely, but it usually includes a drainage supporting member of perforated metal or well screen wire overlaid by a septum of stainless-steel wire or wire mesh. Cylindrical elements made of flexible woven wire; synthetic mesh; or a porous, ceramic, heat-treated material are available.

FILTER SEPTUM DESIGN

The filter cake forms on and is supported by the septum. The size of the clear openings in the septum must be small enough for the precoat media to form and maintain stable bridges across the openings. A clear opening of 0.005 in. (about 125 µm) or less in one direction is desirable. In stainless steel, a 24×110 Dutch-weave wire mesh has been successfully used with the grades of precoat media found in water filtration. An air permeability rating of 70–100 scfm/ft^2 at a pressure drop of 0.018 psi (0.5 in. of water) in association with clear openings of 0.005 in. indicates a satisfactory septum material.

Finer mesh septum material may be desirable if finer grades of filter media are required. For example, a 50×250 wire mesh having a 60-µm retention is available.

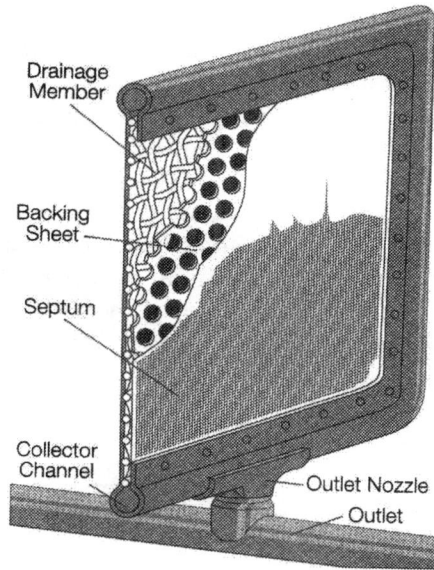

Reprinted with permission of Celite Corporation.

Figure 3-8 Typical construction of a flat leaf filter element

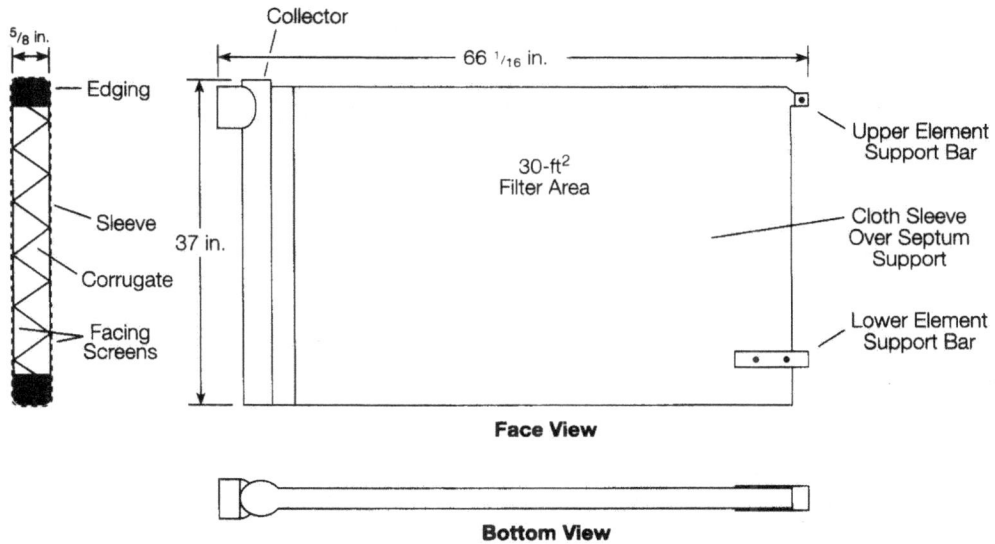

Courtesy of Westfall Manufacturing Company.

Figure 3-9 Typical construction of vacuum DE leaf filter element

The finer mesh screen exhibits a higher pressure drop than the 24 x 110 Dutch-weave wire mesh noted previously. Where possible, evaluations of different mesh screens should be included in pilot studies.

The septum should be firmly supported so that it does not yield, flex, or become distorted as the differential pressure drop increases during the filter cycle. If the septum yields or gives as pressure increases, the filter media bridges may slowly break down. When this occurs, small amounts of filter media may enter the finished water, increasing turbidity. It is also important for the septum to allow any particles that have passed through the filter media also to pass through the septum itself, or else the septum will become plugged and require cleaning. A septum made of monofilament material having uniform, consistent opening sizes resists plugging better than multifilament materials or those that depend on labyrinth passageways to retain the filter media.

HYDRAULICS

A basic criterion in design is to ensure adequate flow velocities within piping and filters to transport the source water and filter media particles to all parts of the filter. This is particularly important when the system is operating at minimum design flow. For most filters, sufficient upward velocity should be maintained to keep particles in suspension and moving. All filter media and dirt particles should become lodged on the element rather than on the bottom of the filter.

Appropriate flow distribution or baffling within the filter vessel should be provided to distribute flow evenly and to prevent short circuiting or scouring of the filter media cake during filter operations. Proper spacing of the filter elements is necessary to provide for cake accumulation, to permit adequate cleaning, and to maintain good hydraulics within the filter.

The drainage characteristics of the filter vessel are important for filter systems in which the spent filter media cake is washed away from the septum and removed from the filter vessel as a slurry. The filter vessel drain should be large enough to discharge the water slurry quickly. The bottom of the vessel should slope to the drain, or the vessel should be equipped with internal flush headers to move all of the solids to the drain. Clean water flush nozzles should also be provided so that the interior of the vessel can be flushed clean of all solids. Any solids remaining in the filter vessel after cleaning could be resuspended and deposited on the clean septum during the next precoat cycle, leading to eventual plugging of the septum.

FILTRATION RATE

As described previously, filtration rate is defined in terms of the volume of liquid that passes through a given area of filter in a specific time. The rate used for design purposes is most frequently an economic judgment based on a number of factors. In the past, a typical filtration rate of 1.0 gpm/ft^2 has been used for design; however, studies involving full-scale equipment have shown filtration rates of up to 3.0 gprn/ft^2 to be most frequently used in treatment plants.

The actual filtration rate in service may need to be varied because of normal variations in demand for water. It is important, however, that the filters not be designed for direct operation according to system demand. Increases or decreases in filtration rate within practical limits do not significantly affect the quality of the finished water, so long as the needed changes are made slowly to avoid hydraulic shock to the developing filter cake. Most of the source water particulate are removed on or near the surface of the filter cake, and removed materials are not normally dislodged by gradual rate changes. Any filtration rate change should include a gradual change in the rate of body-feed to maintain the body-feed proportion to the filtration rate.

DESIGN CAPACITY

The size of the filters and the auxiliary pumps and piping can be determined based on the desired total design capacity. For preliminary design purposes, a common filtration rate for sizing precoat filters for potable water is 1.0 gpm/ft^2. It is essential to carry out suitable field pilot studies to determine the actual design criteria required for effective filtration of the source water. For example, full-scale pilot studies conducted for New York City have shown filtration rates from 2–3 gpm/ft^2 to be practical for Croton Reservoir water treatment.

Filter production is normally controlled by water levels in a clearwell, by levels in elevated storage tanks, or by pressure measurements in the distribution system. Filter operation is normally "on" at a constant flow rate, which can be preset at different rates.

When finished water demands to the system are satisfied, the filter may be placed on "hold" (a recirculation mode) to hold the filter cake in place. In this way, the filter will be ready to go back into production immediately when demand so requires, and usable filter media will not be wasted.

Adequate storage should be available to handle hourly demand fluctuations so that the filter system can be designed to produce treated water at the maximum-day demand rate. Emergency increases in flow rates through the filters may be made within the limits of the treatment system without appreciable deterioration of finished water quality. However, increased loading rates will cause shorter filter run lengths.

A decision to use single- or multiple-filter units depends on several factors:

- regulatory requirements
- finished water storage capacity
- distribution system demand characteristics
- turnaround time of the filter
- availability of alternative water supplies
- availability of spare parts
- overall safety factor desired
- economics

Multiple filters with accompanying equipment should be considered for municipal water supply systems. Systems are often designed to meet maximum-day demand with one unit out of service. Multiple units have several advantages. They will

- allow production to be matched with demand
- provide operating flexibility
- lessen the impact of equipment being out of service during repair or maintenance

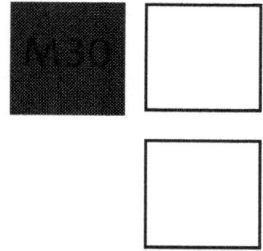

Chapter **4**

Filter Media

Filter media is the basic means for removing contaminants from water in precoat filtration. The media must capture particulate matter and prevent break-through throughout the filter cycle. At the same time, the media must be porous and remain so to permit the water to pass through the cake.

Selection of media appropriate to the quality of the source water and the desired quality of the finished water is an important step in designing an economical precoat filtration system. Equipment and procedures for handling the media must also be specified.

MEDIA TYPES AND GRADES

Diatomaceous earth (DE or diatomite) and perlite are both used as filter media for water precoat filtration. Both materials are available in a variety of grades, allowing the selection of a filter cake with the desired pore size. Various manufacturers make various grades of media, and until recently there has been no uniform method of defining grades.

The American Water Works Association's Standards Committee on Filtering Materials has produced AWWA B101-94, Standard for Precoat Filter Materials, which covers quality control, density, permeability, and particle size distribution for precoat media. The standard should. help both operations personnel and manufacturers define and evaluate diatomite for use in drinking water filtration plants.

Diatomaceous Earth

Diatomaceous earth is the most common filter media used in precoat filtration of water. It is composed of fossilized skeletons of microscopic water plants called diatoms. Diatomite is almost pure silica. It is mined worldwide. Deposits in Lompoc, Calif., are believed to be the largest marine diatomite deposits in the world. Large deposits of freshwater diatomite also are found in Nevada.

Processing the crude DE into a substance suitable for use as a filter media requires milling, calcining, and/or flux calcining (a high-heat process), as well as air classifying into various grades. The processing steps, particularly the flux calcining, cause both physical and chemical changes in the diatomaceous earth to increase its permeability and purity.

Diatomite grades normally used for potable water filtration have a mean pore diameter ranging from about 7 to 17 μm, with median particle size ranging from 23 to 36 pm. Each grade has its own range and distribution of particle sizes. Wet diatomaceous earth has a specific gravity of about 2.3. Unless a diatomaceous earth slurry is kept moving either in the slurry tank or in feed lines, it will settle out and plug feed lines or low spots. Design engineers should be familiar with the unique properties of diatomaceous earth in order to design successful precoat filtration facilities.

Perlite

Another material used for precoat filtration is perlite, which is derived from glassy volcanic rock. The ore is crushed, subjected to high heat (calcined), milled, and classified into several grades. Some of the particles remain as small glassy spheres, which may float and are therefore ineffective as filter media. The dry perlite is less dense than dry DE, weighing from 32 to 70 percent as much as DE. The value of the lighter density material should be proven by pilot studies (for proposed plants) or full-scale studies in existing plants using precoat media. Perlite's lower settling rate makes it less likely than DE to migrate to the lower portions of the filter.

Perlite has been used less frequently for potable water production than DE. However, water treatment plants in Rawlins, Mills (near Casper), and Saratoga, Wyo., have been using perlite successfully for over a decade.

Media Grades

The filter media grade is selected according to the desired flow rate and quality of the finished water. Media grade refers to the median pore size of the established precoat rather than the quality of the media. Pilot testing can determine two or three grades that will produce a desired quality of treated water, simplifying the task of optimizing a final selection when a full-scale plant is being used. As a starting point, a rule of thumb is to match the cake median pore diameter to the median particle size of the material being removed. Selection of a coarser or finer grade of media is dictated by a compromise in terms of quality of finished water versus flow rate within overall economic considerations. When purchasing media, the user should verify that the grades purchased are uniform and consistent.

PRECOATING

The precoat serves two basic purposes. First, it provides an initial filtering surface to trap dirt particles when the filter begins the filtration cycle. Second, it protects the septum from becoming plugged with the suspended solids in the source water. It also acts as a separating agent during cleanup operations.

Precoat Principles

Successful precoating requires that the filter media be applied uniformly to the entire surface of the clean septa. This is accomplished by recirculating a concentrated slurry of clean water and filter media (generally 10 percent or greater) through the filter at 1.0–1.5 gpm/ft^2 until the media is deposited on the septa. Turbidity of the recirculating water is thereby lowered to the desired finished water quality. Precoat recirculation may be a timed cycle based on operating experience, or it may be controlled by turbidimeter output signals.

As the media particles crowd together when passing through the openings, they jam and interlock, forming a bridge over the openings. Because the particles are much smaller than the clear openings in the septum, they are retained and form a stable precoat as a result of this bridging. As the bridges form, additional particles are caught and the filter septum is coated.

The amount of media used for precoating should be adequate to cover the surface of the septum. It typically varies from 0.15 to 0.20 lb/ft^2 of filter area. The thickness of coating will generally be 1/16 to 1/8 in.

The time required to complete the precoat step depends on several factors:

- concentration of the slurry
- the grade of filter aid
- the weave of the filter septum
- the flow rate of recirculation

A septum with small clear openings will form bridges more quickly and complete precoating in a shorter time. However, it may not provide optimal filtration operation since the finer filter cloth or wire mesh may be subject to blinding or plugging.

Precoat Equipment

The most common precoating systems (see Figures 1-2 and 1-3 on pages 3 and 4) require a precoat slurry mix tank and, for pressure filter systems, a precoat recycle tank (see Figure 1-2). The precoat recycle tank is generally not required when a vacuum filter system is involved. In vacuum filter systems, the precoat slurry from the slurry mix tank can be pumped or educted directly into the open filter inlet piping. The precoat system requires the necessary slurry-transfer pumps or eductors, slow-speed mixer, and associated valves and piping connecting the tank with filtered water and the filtration vessel.

Precoat mix tank. The precoat mix tank is sized to hold a volume of 10–12 percent (by weight) concentration of filter media and water, which will provide one or more precoats. For smaller systems with a single filter, it may be advisable to provide tank capacity for only a single precoat to avoid possible degradation of the precoat material from long periods of mixing. To obtain a 1/16 to 1/8-in. precoat, the amount of precoat required will vary from 0.15 to 0.20 lb/ft^2 of filter area. The precoat mix tank should be equipped with a slow-speed mixer (40 to 60 rpm); a sight glass or other means for determining slurry level; effective dust control; and necessary fill, drain, and overflow lines. The tank should have a dished or equal bottom to facilitate cleaning and flushing.

When precoating begins, the filter and recirculation piping are filled with filtered water and recirculation is started. In a pressure filter system, the precoat slurry is then pumped or educted into the recycle tank. In the case of vacuum filters, the slurry goes directly into the inlet piping to the filter or filters. Small, open-impeller, centrifugal pumps or eductors are used to transfer the proper volume of precoat slurry. The eductor, when used, must be supplied with filtered water for motive power. Pumps with mechanical seals are desirable, and pump rotational speeds of 1,750 rpm or less have proven least harmful to the media.

The required slurry volume should be transferred as rapidly as practical. The amount of slurry removed is normally controlled by level probes or timers.

Fill/precoat recycle tank. Until fail-safe backflow-prevention devices became available, large open tanks stored filtered water for use when the filter system was filled in preparation for precoating. Such a tank held about 125 percent of the liquid volume of the filter and the piping related to the precoating operation. It was filled with plant-filtered water piped through an air gap or break. After the system was filled, the liquid remaining in the tank was usually sufficient to permit rapid turnover of the liquid during the precoating operation and to provide flooded conditions on the suction line of the pump used for recycling. The return line from the filter to the precoat recycle tank extended below the surface of the recirculating water in this tank, thus providing sufficient agitation to keep the precoat media in suspension and to avoid air entrainment. A mixer was required if the filter media could not be kept in suspension by the turbulence of the returning liquid.

With proper use of backflow-prevention devices, though, the filter and connecting piping can now be filled directly from the plant potable water system with the result that a much smaller precoat recycle tank can be used.

Regardless of the method used to fill the filter, it is important that all air in the pressure filter be vented during filling to ensure the filter is completely filled and that all filter elements are submerged in the liquid.

In most installations, the raw water pump (for pressure filters) or filtered water pump (for vacuum filters) is piped and valved to serve also as the precoat recycling pump. Where necessary, a separate pump is used for this service. Systems not requiring a large fill/recycle tank can use a smaller tank sized to handle only recycle flow. The prepared precoat slurry is added directly to the water in the open tank, and recycling is continued until the filter media has been deposited on the filtering surfaces and recycle water turbidity meets finished water quality requirements.

The fill/recycle tank and/or recycle-only tank should have a dished or equal bottom and suitably located drains for ease of cleaning.

All piping used for transport of slurry should be sized to permit turbulent velocity at the designed pumping rate and should be as free as possible of sharp, right-angled bends. Tees, rather than bends, provide cleanout access, if required. Cast-iron, steel, and plastic piping and clean, flexible tubing have been used successfully in this application.

BODY-FEED

During filtration, cake permeability is maintained by adding small, predetermined amounts of filter media to the incoming source water before it enters the filter. This filter media is called the body-feed. The body-feed mixes with dirt particles in the source water and is deposited on the filter septa. As filtration proceeds, the cake deposited on the septa is a mixture of filter media and suspended solids removed from the source water. The cake gradually increases in thickness, and although the cake remains permeable so that the water passes through it, the differential pressure across the cake gradually increases because of the increased thickness and resistance of the cake. Eventually, the differential pressure reaches a maximum limit, and at that point the filter cycle is terminated.

Body-feed Principles

The amount of body-feed to be added to the source water is determined by the nature and amount of solids to be removed. Pilot testing during representative source water quality periods will generally indicate the type and range of solids that will be encountered and the amount and type of body-feed needed for efficient operation. As a rule of thumb, nondeformable solids require a body-feed of 1 mg/L of diatomite for each 1 mg/L of suspended solids. Deformable solids may require up to 10 mg/L or more of diatomite for each 1 mg/L of suspended solids. Filter media other than diatomite may require more or less than these amounts.

Proper control of the body-feed system and its rate of application is the most important factor contributing to economic operation of a precoat filtration plant. Accuracy of feed rate and continuity of feed are critical to the operation. Feeders should be designed to maintain accuracy over a wide range of feed rates so they can be closely matched to varying source water conditions. Plants may be equipped with instrumentation so that body-feed rates automatically adjust to changing source water turbidity. Interruptions in body-feed should be avoided as much as possible to lessen the chances of premature filter cake blinding. Any sudden increase in pressure drop that is associated with loss of body-feed cannot be offset by subsequent higher body-feed rates. The net result is that the filter cycle will be shortened.

Body-feed Equipment

Proper design of body-feed metering equipment is critical for good system performance. Inadequate design in this area has been the principal cause of difficulty with precoat filtration systems in the past. Body-feed equipment can be classified as dry or wet systems.

Dry feeders. Most dry feeders are similar in design to the feeders used for lime or other dry chemicals in the water utility industry. In these units, the dry chemical is fed directly into the incoming source water from storage hoppers that incorporate a metering device. Because some filter media tend to bridge readily, a vibrator or other means to ensure movement of the dry material should be used when feeding from a hopper. The feed equipment should be protected from excess moisture because filter media may absorb moisture from the air, which could further complicate dry feeding.

Where the dry chemical cannot be metered directly from the hopper to the incoming source water, it can be fed into a small slurrying device, such as an eductor, and then introduced to the source water inlet flow to the filter. Untreated water may, in most cases, be used for operation of the eductor. Because of the relatively large volume of water required for eductor operation, use of an eductor in small systems needs careful evaluation. This is particularly important if finished plant water will be used for the eductor, since the possible excessive dilution of the source water could pose operating problems due to fluctuations in source water quality.

Dry feeding reduces the space needed for equipment and storage, which could be important in large plants.

Wet feeders. Probably the most common devices for providing body-feed are wet feeders. These generally include a slurry mixing tank, mixer, and a metering pump to deliver controlled amounts of the slurry to the source water. The pumps are usually designed to handle slurries of less than 10–12 percent concentration. However, better overall performance can be expected if the slurry concentration used is much lower. The metering pump should be certified by the manufacturer for use with diatomite or perlite. Performance curves for the slurry concentrations to be handled should be provided by the manufacturer as well.

Filter media is relatively abrasive and settles out rapidly from slow-moving liquids. Certain designs are recommended to minimize problems in operation. Ball check valves with hardened steel balls and wear-resistant elastomeric seats are suggested for the design of body-feed systems using pumps. In addition, pumps should be equipped with flush connections so that they can be flushed periodically with either source or filtered water to minimize potential plugging, or they should be provided with cleanout fittings.

Provisions to prevent backflow of flushing water into the slurry tank are necessary. The body-feed lines carrying the slurry should pitch down from the pump discharge to the filters source water feed line and should be as short as possible. Vertical discharge piping of more than 1–2 ft should be avoided to minimize deposition of body-feed on the pump discharge valve or valves. Flexible transparent tubing instead of pipe will allow any clogging to be easily located and dislodged. Body-feed pipe or tubing should be sized to maintain turbulent velocities to keep the media in suspension. Before extended shutdown periods, all slurry lines should be flushed.

FILTER MEDIA STORAGE

Filter media is normally supplied in bags, except at large plants, where bulk storage may be advantageous. Precoat filter media is packed in multiwall paper bags containing approximately 2.1 ft^3 of media, wrapped and shipped in plastic on pallets for storage and handling. Also available are semibulk bags weighing 1,000 lb, which may be unloaded directly into a large-capacity slurry mix tank or incrementally in exact amounts using an unloading stand equipped with volumetric or gravimetric feeders. Material-handling equipment for moving loaded pallets will be needed. Plant and equipment layout should facilitate moving bags from the storage area to point of use. Plants should have adequate storage for at least 30 days of operation.

Dust collection systems must be provided when bags are unloaded into dry or slurry feed tanks. Anyone handling dry filter media should wear an appropriate respirator to prevent dust inhalation. Dry storage hoppers should be arranged so as to minimize moisture pickup, and they should be equipped with dust-control systems. Wet-slurry systems should incorporate slow-speed (40- to 60-rpm) mixers that will maintain the media in suspension but will not degrade the media particles by excessive agitation. Because the media is difficult to resuspend, mixing should be continuous so that the material does not settle.

Bulk storage systems may be appropriate and economical for larger plants (those using 100 tons per year or more) because of savings from reduced storage space, bulk purchases, and reduced labor for bag handling.

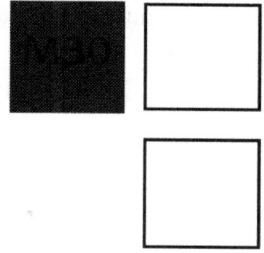

Chapter **5**

Spent Cake Removal and Residuals Handling

For a precoat filter to operate effectively over an extended period, the accumulated filter media/dirt cake must be fully removed at the end of each filter cycle, producing a clean septum for the next precoat. The waste material resulting from the filter operation, both backwash water and solids, must be disposed of in an acceptable, economical reamer.

Regulations governing the disposal of water treatment plant residuals have become more strict. The new regulations affect not only the allowable disposal practices, but also the cost of disposal in accepted landfills. Thus, thorough cleaning using the least backwash possible and minimization of water treatment waste are of growing importance to efficient operation of a water treatment plant.

SPENT CAKE REMOVAL AND CLEANING METHODS

At the end of each filter cycle, the spent filter cake is removed from the filter in preparation for the precoat that will begin the next cycle. If spent solids are not fully removed from the filter vessel, the material could be resuspended and deposited on the filter septum and eventually become a problem. The effect would usually develop gradually so that the operator would become aware of it only after a number of cycles.

The most reliable determination of septum cleanliness requires visual inspection of the bare septum, followed by observation of the uniformity and completeness of the precoat. When elements cannot be fully inspected, a higher-than-normal differential pressure immediately after precoating (at the start of the filtration cycle) might mean that the septa are becoming fouled. A lumpy precoat or bare spots on the septum to which precoat will not adhere are signs of possible septum plugging.

Techniques for cleaning filters vary with the different kinds of filter vessels and filter elements. The most common methods involve sluicing; flow reversal; or draining, drying, and vibrating.

Sluicing

In this process, the cake is removed from the elements with high-pressure internal sprays directed onto the exterior surface of the element.

In sluice cleaning, the entire septum area should be covered by either stationary or oscillating fan spray nozzles that deliver water under a minimum pressure of 60 psig, with sufficient flow to cut away and flush the spent cake from the septa. The nozzles should be designed so as not to become plugged with solids during filtration.

Some filters have the capacity to rotate the filter element under the backwash spray. A filter design that provides this capacity needs sufficient room for the rotating filter leaves to clear the spent diatomaceous earth (DE or diatomite) at the bottom of the tank before it washes to drain.

A good maintenance practice is to clean the septa at least once a year or as needed with a high-pressure wand, or to scrub the septa. Care must be exercised if scrubbing is practiced since the fine synthetic on metal-cloth fibers is easily damaged.

Flow Reversal

When flow reversal is used to clean the filter elements, the velocity and volume of flow must be adequate to dislodge all the spent cake from the septa. In some cases, an air bump technique is used in which a volume of high-pressure air is trapped within the vessel. A drain valve is then quickly opened, and the sudden release of pressure causes a momentary high reverse velocity (a blasting effect) that dislodges the cake. In equipment using this flow reversal method of cleaning, the cylindrical septa, which are long in relation to diameter, are generally clustered close together with small spaces between the septa. There is a possibility the blasting effect will cause the spent cake to bridge between septa, so a sufficient flow of water must be used to flush away the bridges.

Draining, Drying, and Vibrating

Another method is to drain the tank under differential air pressure, dry the cake, and then vibrate the leaves to dislodge the cake. This method is more commonly used in industries for which the liquid being filtered is quite valuable and must be reclaimed, or where dry cake can be handled more economically than slurry.

SLUDGE AND WASTE HANDLING _____

The spent cake removed from a precoat filter is a mixture of filter media and the materials removed from the influent source water. This waste matter is initially removed from the filter in a slurry form. Although some systems may be able to dispose of the slurry into a sanitary sewer system, most plants must dewater the waste material and make separate provisions for the solid and liquid wastes.

Settling Basins

The most common method of handling waste solids and water from precoat filter operations is to discharge it into settling basins. Filter media settles rapidly, and complete settling is typically accomplished in less than one hour. This permits the backwash water to be recycled or to be discharged if it satisfies regulatory requirements. Regulatory requirements may differ depending on whether the backwash water discharge is to a natural body of water or to the sanitary sewer system. Ultimately, the basins will fill with the solids and the solids will need to be removed to a landfill or other disposal site.

Settling basins should be sized to contain the total volume of backwash water from at least two filter cleanings, plus the volume of solids that are expected to accumulate in the time between basin cleanings. Collection and characterization of backwash samples should be conducted during pilot studies to aid in the sizing of collection basins for final design. More than one settling basin should be provided so that one basin can receive waste from the plant while another is being cleaned. When designing the system, engineers should consider the method of cleaning to be used, as well as access for people, backhoes, loaders, or other means of removing spent precoat material.

An overflow weir is often a sufficient outlet for the basins. However, a decanting system to optimize the withdrawal of clarified supernatant may be preferable. A means for dewatering the basins should be provided so that the maximum degree of natural drainage dewatering will occur before the solids are to be removed.

Mechanical Dewatering

Waste solids from a precoat filter plant are more easily dewatered than solids from conventional coagulation/filtration plants. Precoat filter residuals can typically be dewatered by vacuum filters or belt filter presses. The resulting cake is consistently dry and easily handled. Common practice is to return the liquid filtrate from the mechanical dewatering processes to the source water feed to avoid a liquid disposal problem. However, the quality of the recycle stream should be carefully monitored and provisions made to regulate the proportion of recycled backwash water in the influent to avoid problems with the filtration process.

Final Solids Disposal

The ultimate disposal of retained particles and media solids from precoat filtration plants is on land. The most desirable disposal incorporates a beneficial use for the spent media. Examples of beneficial uses that are being tested are use as a soil conditioner, mixture with sand for sand traps on golf courses, and mixture with sand used for road sanding in areas where winters require such measures. The media material is predominantly inert silica. Mixed with soil, it helps to make the soil loose and friable, a particular benefit to clay soils.

Backwash material should be tested during pilot studies to characterize the turbidity component of the waste solids. Caution should be exercised in the ultimate disposal of the waste solids if the treatment has included oxidation or precipitation of possible toxic components. Care should also be used in disposing of components that are harmful to agricultural uses.

Sanitary Sewers

Under the proper circumstances, waste solids can be disposed of to a sanitary sewer system. This method should be considered only when the solids from the water treatment plant represent a minor percentage of the total solids being handled at the wastewater treatment plant. Caution regarding the possible introduction of toxics or excess loadings is in order for this method of waste disposal as well as for land disposal.

Where this method is used, the waste slurry should be metered into the system slowly in order to prevent surge loading on the wastewater treatment plant. The solids will represent an additional loading on the plant. The solids settle rapidly and will be removed during primary clarification. Sufficient velocity of flow in the sewer piping should be maintained to prevent the solids from settling in the piping before

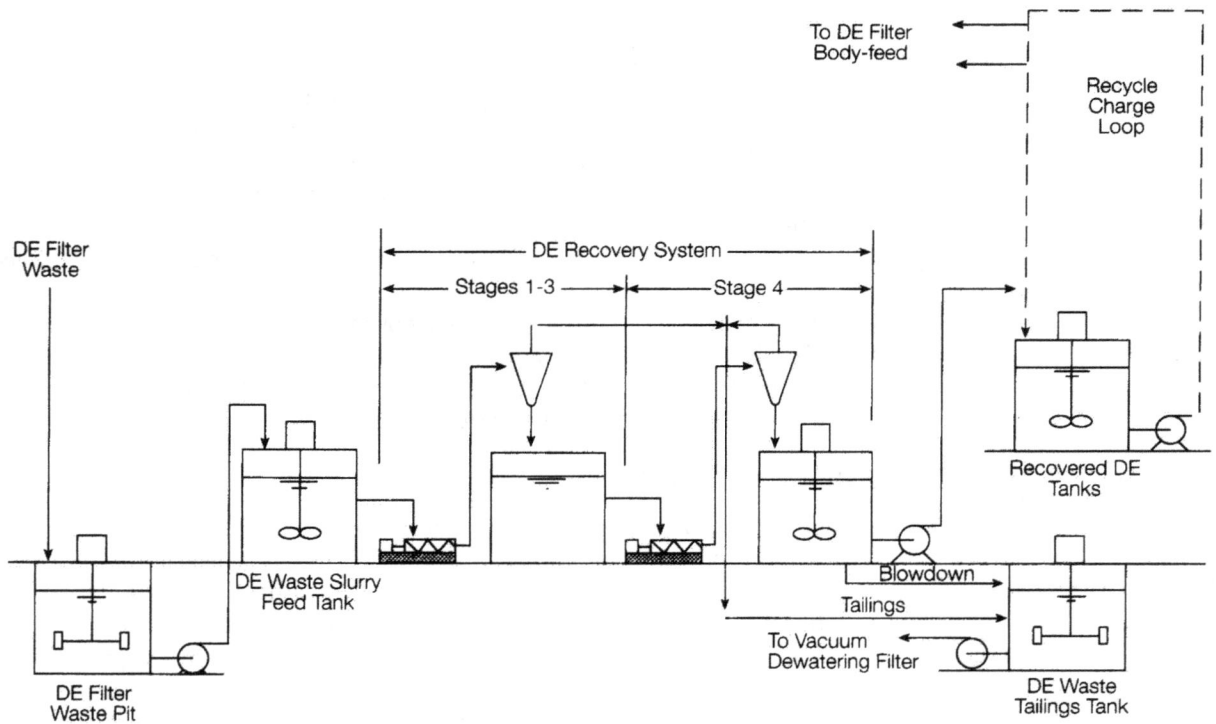

Courtesy of Applications Corporation.

Figure 5-1   Schematic flow diagram of DE recovery system

they reach the wastewater treatment plant. This additional loading on subsequent sludge-handling processes, such as digestion, should be considered.

Recovery and Reuse

Research (Principe et al. 1994) is focusing on the feasibility of recovery and reuse of spent DE. Early research has indicated that up to 90 percent of the DE used in the run may be recovered, cleaned, and reused as a body-feed. Figure 5-1 shows a schematic of a prototype recovery system. Recent full-scale operation of a filter aid recovery system has shown it to be suitable for plants that treat more than 10.0 mgd.

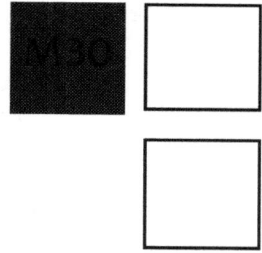

Chapter 6

Operation, Process Control, and Monitoring

Precoat filtration with either diatomaceous earth (DE or diatomite) or perlite as a media is not a complex process. Process control of the precoat filtration technique does not require constant monitoring of solution chemistry to achieve consistent results. Because it is not chemical in nature, as many other water treatment processes are, the process eliminates the need for the operator to have a knowledge of chemistry to understand plant processes.

The amount of operator attention required by a precoat filtration installation varies greatly with the degree of automation designed into the plant and plant capacity. The design engineer's careful application of the precoat filtration process to only those source waters that are of consistent quality and are well suited to this method will greatly simplify the operator's task of ensuring a high-quality finished product. Automation must not detract from the operator's ability to control the quality of the finished product. As with any operation, the primary duties of the operator fall in the areas of daily monitoring and adjustment, preventive and unscheduled maintenance, and record keeping.

OPERATING ADJUSTMENTS

In plants with manual operation, operators are principally involved in regulating and gradually adjusting flow rates so that production matches the need for water. Periodic calibration of body-feed and precoating systems to ensure accuracy of feed rates is necessary. Body-feed and precoat slurries must be prepared and effluent water quality monitored on a regular basis. Operators also need to adjust body-feed rates and the rates of any other chemical feeds needed for raw water preconditioning.

The degree of automation of the precoat filtration process, as with any water treatment method, is limited only by current technology and the imagination of the design engineer. The need to automate appears to be for the operator's convenience,

35

not due to process requirements. The current practice of precoat system operators is to automate as the availability of a qualified operator's time dictates.

In partly or fully automated plants, the operator is primarily concerned with monitoring flow rates, body-feed rates, and effluent water quality. Pressure sensors, level-controlling devices, and rate-of-flow controls normally adjust operations automatically to meet changing demand. The operator should be able to override the automation manually and introduce new control settings that vary operation to meet changing demands.

System operation controlled by a maximum system pressure differential could be automated. As a minimum, however, the cleaning and precoating cycles should not be automated and should require operator intervention to ensure that these operations are performed thoroughly. Manual overrides of all portions of the operation must be provided for maximum operator control and for production flexibility.

Operators are encouraged to establish support networks to provide a forum to discuss system operations and to take advantage of operational experiences at similar precoat filter water treatment plants.

PROCESS CONTROL

The critical parameters of precoat filtration process control are rate of flow and body-feed control. Precoat filters are normally operated at a constant rate of flow to simplify the body-feed control. The body-feed media can be added at a constant rate provided the source water is of consistent quality.

Flow Control

Constant system flow can be accomplished in two ways. The traditional method is to use a constant-speed, motor-driven pump and a mechanical rate-of-flow control valve on the discharge side of the filter. A more efficient method is to use a variable-frequency drive (VFD) on a high-efficiency motor with an electronic flow controller on the filter effluent. The VFD application can result in significant energy savings because the speed of the motor is specific to the pressure change across the filter. The need for cleaning in both cases is controlled by a predetermined differential pressure limit.

Variable-frequency drives may reduce operating costs during periods of low system demand. They provide operators with the ability to operate the main pump at a lower speed specific to the reduced demand, or to recirculate within the filter. Flow through and around the filter (recirculation) must be maintained, even if the filtered water is not delivered to the system. If the filter were to shut down in response to low system demand, the filter cake would fall off the septum. Before variable-frequency drives were available, the older design method was to use a separate fixed-speed recirculating pump to maintain the cake on the septum. Both recirculation methods retain the filter cake on the septum, and the filter is immediately ready to be returned to service when demand for water from the system again signals for filter production.

In precoat filter operation, it is important that all changes in flow be made slowly and smoothly. All flow control valves should open and close gradually without sudden slamming. Quick valve movement introduces sudden pressure pulses or surges that tend to disturb the filter cake, causing the particles that bridge across septum openings to break down slightly and cast off small amounts of media into the finished water. Automatic, electronic operation of valves can easily address this problem.

Body-feed Control

The level of automatic control in body-feed systems is a function of the degree to which the design engineer chooses to utilize current technology. System operation can be as basic as maintaining constant body-feed rate, regardless of influent water quality, and varying by trial and error the length of the filter runs. On the other hand, control of the body-feed rate can be electronically linked to source water quality parameters such as turbidity or iron content. Body-feed rate can also be correlated to the output flow rate. Maximum operating efficiency requires having the option to easily adjust the body-feed rates to variations in system conditions.

Means for controlling and varying body-feed rates must be provided. These controls should be easy to set and adjust; they should be easy to calibrate periodically; and they should maintain accuracy of feed with time.

Normally, control systems will include pumps or other devices that have a variable volume output range, That permits the operator to easily change the volume of body-feed slurry being fed in order to match changes in influent flows or source water quality. If body-feed rates are difficult to change, there is less likelihood that the plant will be operated at optimal conditions.

MONITORING

Pressure, flows, and turbidity are important parameters to monitor in the operation of a precoat filter. Permanent recording of this data is vital. The ability of the operator to observe trends and to forecast performance is invaluable.

Pressure

Inlet and outlet pressure at the filter vessel should be both monitored and recorded by suitable instrumentation. This will provide a recorded history of plant operation, giving important indication of the rate of pressure differential rise during filter cycles, as well as indicating the efficiency of cleaning (as shown by low pressure differentials as filter cycles begin).

Flow

Flow through the unit should be measured, recorded, and totaled. It is helpful to record total water volumes used during filter cleaning and for other internal plant usage. This information can be used to maximize the net amount of water available to serve system demand.

Turbidity

All regulatory agencies recognize finished water turbidity as a valuable parameter for assessing the performance of the filtration process. Instruments for continuous turbidity measurement of the filter effluent should be provided. Source water turbidity also should be monitored as a critical guide to setting body-feed rates. Body-feed pumps or other body-feed systems may be controlled automatically by continuous monitoring of source water turbidity and by feed rate adjustments in proportion to turbidity changes.

Turbidity instrumentation can be designed to sound an alarm, both audible and visible, when finished water quality deteriorates beyond a point set by the operator. It should be designed to shut down the filter or put it into recirculating standby mode until the problem is corrected.

EMERGENCY CONTROLS

Under certain circumstances, temporary changes to normal operating parameters can require that the filter cycle be terminated and restarted.

Interruption of Flow

In precoat filtration, it is critical that a flow of water through the filter cake be maintained at all times until the filtration cycle ends. Even a momentary interruption of flow due to a power outage, pump failure, valve malfunction, or similar cause may disrupt the stability of the cake or dislodge it entirely. Once the integrity of the cake has been damaged, any resumption of flow may allow some unwanted filterable particles to pass through the septa into the finished water.

Use of a time delay device would span any momentary power failure that could result in loss of the filter media. At the very least, the system should have alarms that will alert the operator to any interruption of flow or any change in treated water quality. In such events, operator intervention will be required before the system is restarted. Operators should also monitor loss of power to, or drop in performance of, any ancillary equipment to the filter that could affect proper operation.

Automatic filter shutdown is appropriate in some cases as determined by a preset maximum pressure differential across the filter.

Loss of Air or Auxiliary Motive Power

Any loss of motive power to the devices that control the filter may result in improper operation, and the system should be cleaned and precoated before being placed on-line again.

Alarms

Visible and audible alarms to notify operating personnel of filter operating problems should be provided. The required alarms should alert operators to power failure, rupture-disk failure, high turbidity, and loss of flow or loss of pressure.

MAINTENANCE

Many of the maintenance requirements in a precoat filtration plant are similar to those in other water treatment plants. Standard maintenance practice should be observed for pump, valves, mixers, compressors, and electrical components.

Items that are specific to precoat filtration plants include filter media handling and feeding equipment, the filter vessel, and filter elements. This equipment should be routinely inspected for wear due to abrasive media. Body-feed pumps and lines carrying the slurry to the source water inlet should be inspected daily to detect possible plugging. Daily observation of pressures, flows, and general appearance of the filter cake after precoating, during filtration, and after cleaning will also detect potential maintenance problems. High pressure differential at the start of a filtration cycle and excessive rate of rise in pressure during filtration suggest possible septa plugging and the need for cleaning or maintenance. Unusually low pressure differential or high effluent turbidity could be an indication of a ruptured septum.

RECORD KEEPING

One of the most important duties of the plant operator in running any plant is to routinely and regularly record all pertinent information relating to the overall performance and operation of the plant. State regulatory authorities and the federal Surface Water Treatment Rule require operators to keep water quality records, including daily observations of filtered-water turbidity for plants treating surface water. Instruments that automatically record the pertinent data are preferable, but if they are not available, the operator must keep records for each day of operation.

Daily records of process operating data, such as the amount of filter media used, body-feed rates, total daily throughput, filter run length, filter pressure differentials, and volume of water used for cleaning, should be kept. These records provide the information necessary to optimize the economics of operation, help interpret data that may indicate potential problems, and provide a recorded history for later use.

All power outages should be logged, as well as maintenance and repair items that are related to the filtration operation. Maintenance of concise, easily accessible records of system operation is critical to efficient process management. Simple database or spreadsheet computer software may be used to accomplish this task.

Chapter **7**

Auxiliary Equipment and Safety

In addition to the main filtration and media-handling equipment discussed in the preceding chapters, the operational precoat filtration plant requires various pumps, valves, and other auxiliary equipment. This chapter discusses general requirements for such equipment and treats the issues of safety as applied to precoat filtration operations.

AUXILIARY EQUIPMENT

Auxiliary equipment used in precoat filtration includes pumps, valves, treated water supply systems, and standby equipment. Some of this equipment must be selected to meet the special needs of handling abrasive filter media slurry.

Pumps

In general, standard water utility pump practice should be followed in the selection of pumps for use in precoat filtration plants. Certain precautions should be followed, however, because of the abrasive nature of filter media. These precautions are of particular importance if a pump is to be used for two different steps of the filtration process. This might occur if a pump will be used to apply precoat to the filter, and with a change of valving, also supply source water to the filter. Filter media slurry concentrations of less than 1 percent are not normally harmful to pump impellers or other liquid-handling equipment. However, even small concentrations may cause damage if media particles become embedded in the pump packing. Where practical, mechanical seals should be used or packing should be supplied with clean water from an external source.

When diatomite (DE or diatomaceous earth) or perlite concentrations greater than 1 percent are being handled, some erosion of impellers and other moving contact parts may eventually occur, particularly with high-speed rotation in excess of 1,750 rpm. For these applications, parts made of abrasion-resistant materials,

41

coupled with a low rotation speed, may be appropriate. Consideration should be given to the use of positive displacement pumps for handling diatomite or perlite slurries in concentrations up to 5 percent.

Valves and Piping

In general, valves in precoat plants should be selected in accordance with standard water utility practice, except where high concentrations of filter media slurries are to be handled. Diaphragm valves, plug valves, or butterfly valves should be used in the body-feed or precoat slurry injection systems.

Valves in the precoat recirculation loop can be treated as clean water valves, since they open before the slurry enters the system and close when the precoat recirculation liquid is clean.

A consideration for plant design is to apply the body-feed as close to the filter inlet as practicable to lessen the opportunity for precoat settling in the inlet lines or for valves plugging up ahead of the filter. In the same vein, keeping the number of bends and piping obstructions to a minimum in the precoat piping and body-feed lines will help prevent plugging of the piping. Access to all portions of the feed piping is important for ease of maintenance and operation.

Water for Internal Plant Use

Good practice in precoat filtration operation requires the use of treated water for general filter plant functions. Precoat slurries and water used for filter cleaning should be of the highest quality. Water is commonly drawn from filtered water storage, with appropriate precautions taken concerning cross-connections. In-line strainers may be required on sluice water lines to prevent plugging of the sluice nozzles.

Standby Equipment

Standby equipment should be provided as required by state or local regulations and suggested by sound engineering judgment. Standard practice should guide decisions with respect to pumps, valves, and motors. For precoat plants specifically, however, consideration must be given to dealing with the cleaning or replacement of filter elements in case the septa become plugged or damaged. Generally, plugging of septa occurs slowly, so experience will indicate how frequently they should be cleaned, and planned outages for cleaning can be integrated with low demand or other maintenance operations. It is advisable to maintain several extra filter elements on hand as replacements for those taken out of service for cleaning or repair.

SAFETY

Safety concerns in a precoat filtration plant may include the following:
- handling of the dry filter media by operating personnel
- pressure-tank pressure relief
- pressure-tank opening procedures
- automatic plant shutdown in the event of power failure or other problems

Filter Media Handling

Filter media materials are very abrasive and may cause irritation or injury if incorrectly handled. Diatomaceous earth contains crystalline silica, which has been classified as a probable cause of cancer if inhaled. Perlite may contain crystalline silica ranging from O to 1 percent. Dust collectors must be provided to reduce the fine dust created wherever dry filter media is manually loaded into hoppers or tanks. In addition, a dust collector system is needed above any slurry preparation system. Personnel handling the filter media should be supplied with appropriate respirators, gloves, and goggles, and an eyewash should be available. Personnel should be cautioned not to rub their eyes and to use the eyewash in case of irritation. Coveralls or rubber aprons for operators are recommended. Safety precautions for a specific product should be followed according to the material safety data sheets provided by the supplier.

Slurry preparation facilities must have sufficient clean makeup water provided in the design. Operations personnel should be urged to mix dry filter media into moving water to reduce dust from media. Water movement may be generated by an eductor or by a slow rotation mixer on a slurry tank. The media should be thoroughly wetted without being allowed to settle.

Pressure-Tank Relief Devices

If pressure filters are used, each tank must be equipped with pressure relief devices in accordance with the requirements of the American Society of Mechanical Engineers' Boiler and Pressure Vessel Code, Section VIII. If a rupture disk is used, drain piping must be provided that is capable of withstanding the initial pressure surge and the impact of shattered disk material. The tank overflow resulting from a failure should be carried by the drain piping to a safe location away from operating personnel, electrical equipment, and dry filter media storage. Use of pressure relief valves may be desirable in parallel with the rupture disk.

Pressure-Tank Opening Procedures

Personnel must check that pressure filter tanks are unpressurized before opening a tank for cleaning or inspection. Some manufacturers of pressure vessels incorporate safety features on the access doors to prevent opening unless pressures are equal on both sides of the door. This ensures that the vessel is unpressurized and drained before access is permitted.

Automatic Shutdown

Automatic shutdown of the filtering system should be provided in the event of a power failure or pressure-tank rupture-disk failure. The plant should be brought back on-line manually.

Glossary

amorphous Noncrystalline; having no ordered molecular structure.

area The surface available in a filter for the passage of liquid and formation of a filter cake; usually measured in square feet (ft^2).

backwash A reverse flow of liquid to remove solids from the filter.

baffle A plate or deflector to provide flow distribution in a filter; primary function is to prevent erosion of precoat and settling of body-feed in the filter tank.

blinding Plugging or sealing of any portion of a filter septum by solids that are not removed during the normal cleaning cycle.

blind spots Any place on a filter septum where liquid cannot flow through because of blinding.

body-feed Filter media that is continuously added to the filter that is in operation; purpose is to maintain a permeable filter cake.

cake The accumulation of solids, including filter media and material removed from the source water on the surface of a precoat or septum.

cake space The volumetric space available in a filter to support the formation of a cake.

clarity Clearness of liquid as measured by a variety of methods.

cloth A type of woven filter septum made from natural or synthetic yarns.

colloid Very small, insoluble, nondiffusible solid particles that remain in suspension in a surrounding liquid.

compressibility Degree of physical change in suspended solids or filter cake that is subjected to pressure.

cycle The length of time a filter is in operation before cleaning is needed; frequently used to include cleaning time as well.

DE See *diatomaceous earth*.

deformable Used to describe suspended solids that extrude into the interstices of a filter cake and cause rapid filter plugging.

diatomaceous earth The fossilized skeletons of minute, prehistoric, aquatic plants.

diatomite See *diatornaceous earth*.

differential pressure The difference in pressure between the upstream and downstream sides of a filter cake.

element Any structural member in a filter on which the septum is supported; may be round, rectangular, or cylindrical.

filter aid Any material that helps to separate solids from liquids; frequently used in difficult filtration applications.

filter medium Permeable material that separates particles from fluid passing through.

filtrate Liquid that has passed through the filter media.

filtration The process by which solid particles are separated from a liquid when the liquid is passed through a permeable material.

filtration rate The volume of liquid that passes through a given area of filter in a specific time; usually expressed in terms of gallons per minute per square foot (gpm/ft^2).

flow rate The rate at which a liquid is passed through a system; usually expressed in terms of gallons per minute (gpm) or gallons per hour (gph).

gelatinous   Used to describe suspended solids that are slimy and deformable; capable of causing rapid filter plugging.

interstices   The void spaces in and around solid particles that are packed together.

leaf   Any flat filter element that has or supports the filter septum.

media   The material that performs the actual separation of solids from liquids; sometimes erroneously used to mean *septum*.

mesh   (1) Number of strands in a lineal inch of woven filter fabric. (2) A commonly used synonym for *septum,* as in wire mesh.

monofilament   A single synthetic fiber of continuous length; used in weaving filter cloths.

multifilament   A number of continuous fiber strands that are twisted together to form a yarn; used in weaving filter cloths.

negative pressure   The force of vacuum or suction; usually measured in inches of mercury (in. Hg).

nephelometric turbidity unit (ntu)   A measurement of turbidity in a liquid.

particulate   A substance made up of minute separate particles.

perlite   A form of volcanic rock that, when processed, yields various grades of filter media; also, the filter media so obtained.

permeability   The property of the filter media that permits a fluid to pass through under the influence of a pressure differential.

plate   Any flat-surface filter element; usually found in horizontal plant filters.

precipitate   A reaction forming an insoluble solid that can be separated from the liquid phase.

precoat   The initial layer of filter media that is deposited on the filter septum; usually 1/16–1/8 in. thick on pressure filters.

screen   A term commonly used for septum.

septum   Any permeable material that supports the filter media.

slurry   Any liquid containing suspended solids.

turbidity   An optical measurement of reflected light in a liquid; often used to refer to the particles that cause turbidity.

Additional Sources of Information

Bell, G.R. 1965. Removal of Manganese by Controlled Precipitation and Filtration. *Jour. AWWA,* 57(5):655.

Bryant, E.A., G.P. Fulton, and G.C. Budd. 1992. Improving Performance of Existing Baseline Facilities. Chapter 12 in *Disinfection Alternatives for Safe Drinking Water.* Hazen and Sawyer. New York: Van Nostrand Reinhold.

Bryant, E.A., and C.Y. Yapijakis. 1977. Ozonation-Diatomite Filtration Removes Color and Turbidity, Parts 1 and 2. *Wtr. & Sewage Works,* Sept. and Oct.

Carter, J. 1995. Personal communication. Feb. 14. Mills, Wyo.: Mills-Wardwell Water Treatment Plant.

Cleasby, J.L. 1990. Filtration. Chapter 8 in *Water Quality and Treatment,* 4th ed. Edited by Frederick W. Pontius. New York: McGraw-Hill, Inc., and the American Water Works Association.

Coogan, G.J. 1962. Diatomite Filtration for Removal of Iron and Manganese. *Jour. AWWA,* 54(12):1507.

Dillingham, J.H., J.L. Cleasby, and E.R. Baumann. 1966. Optimum Design and Operation of Diatomite Filtration Plants. *Jour. AWWA,* 58(6):657.

———. 1967. Diatomite Filtration Equations for Various Septa. ASCE *Jour. San. Engrg. Div.,* 93(SA1):41 (Feb.).

Fourth Quarterly Cost Report. 1994. *Engrg. News-Record,* 233(25):33–35 (Dec. 19).

Graham, L. 1995. Personal communication. Feb. 14. Rawlins, Wyo.: City of Rawlins Water Treatment Plant.

Hunter, J.V., G.R. Bell, and C.N. Henderson. 1966. Coliform Organism Removals by Diatomite Filtration. *Jour. AWWA,* 58(9):1160.

Lange, K. P., W.D. Bellamy, and D.W. Hendricks. 1984. Filtration of *Giardia* Cysts and Other Substances. In *Vol. 1: Diatomaceous Earth Filtration.* USEPA 1600/S2-84/114. Sept.

Lawrence, C.H. 1964. Design Aspects of Lompoc Water Treatment Plant. ASCE *Jour. San. Engrg. Div.,* 90(SA6):81 (Dec.).

Logsdon, G. S., J.M. Symons, R.L. Hoye Jr., and M.M. Arozarena. 1981. Alternative Filtration Methods for Removal of *Giardia* Cysts and Cyst Models. *Jour. AWWA,* 73(2):111–118.

Logsdon, G. S., T.J. Sorg, and R.M. Clark. 1990. Capability and Cost of Treatment Technologies for Small Systems. *Jour. AWWA,* 82(6):60.

Manville Filtration Manual—Filter Septa, Section IX. 1991. Manville Filtration and Minerals Group (H.G. Walton).

O'Connor, J. T., and B.E. Benson. 1970. Iron Removal Using Magnesium Oxide. ASCE *Jour. San. Engrg. Div.,* 90(SA6):1335 (Dec.).

Pittendreigh, L.M. 1966. Recent Experiences with Diatomite Filters for Iron Removal on Potable Water. *Johns Manville Wtr. Filt. Sem.,* Boston, Mass. April 14.

Prediction of Diatomite Filter Cake Resistance. 1967. ASCE *Jour. San. Engrg. Div.,* 93(SA1):57 (Feb.).

Principe, M., R. Mastronardi. D. Brailey, D. Nickels. and G. Fulton. 1994. New York City's First Water Filtration Plant. In *Proc. AWWA Ann. Conf.— Water Quality.* Denver, Colo.: American Water Works Association.

Ris, J.L. 1987. Precoat Filtration: Pilot and Full-Scale Design and Operations. In *Proc. AWWA Seminar on Coagulation and Filtration: Pilot to Full Scale.* Denver, Colo.: American Water Works Association.

Ris, J. L., I.A. Cooper, and W.R. Goddard. 1984. Pilot Testing and Predesign of Two Water Treatment Processes for Removal of *Giardia lamblia* in Palisade, Colo. In *Proc. AWWA Ann. Conf.* Denver, Colo.: American Water Works Association.

Schuler, P. F., and M.M. Ghosh. 1990. Diatomaceous Earth Filtration of Cysts and Other Particulate Using Chemical Additives. *Jour. AWWA,* 82(12):67–75.

———. 1991. Slow Sand and Diatomaceous Earth Filtration of Cysts and Other Particles. *Water Research,* 225(8):995–1005.

Spencer, C.M. 1991. Modification of Precoat Filters with Crushed Granular Activated Carbon and Anionic Resins to Improve Organic Precursor Removals. Master's thesis, University of New Hampshire.

Spencer, C.M., and M.R. Collins. 1990. Modifications of Precoat Filters with Crushed Granular Activated Carbon and Anionic Resin Improves Organic Precursor Removal. In *Proc. AWWA Ann. Conf.* Denver, Colo.: American Water Works Association.

Stumm, W. 1977. Chemical Interaction in Particle Separation. *Env. Sci. & Tech.,* 11:1066.

Syrotynski, S., and D. Stone. 1975. Microscreening and Diatomite Filtration. *Jour. AWWA,* 67(10):545.

Task Group Report. 1965. Diatomite Filters for Municipal Use. *Jour. AWWA,* 57(2):157.

Walton, H.G. 1977. Laboratory and Filter for Diatomite Filtration Test. Paper presented at the 43rd Ann. Amer. Sot. of Brewing Chemists Meeting, May, at St. Louis, Mo.

Index

Note: An *f.* following a page number refers to a figure; a *t.* refers to a table.